SAILING YACHT DESIGN: TH

SAILING YACHT DESIGN
THEORY

EDITED BY
CLAUGHTON, WELLICOME & SHENOI

PRODUCED IN ASSOCIATION WITH

Addison Wesley Longman Limited
Edinburgh Gate, Harlow
Essex CM20 2JE, England
and Associated Companies throughout the world

© Addison Wesley Longman Limited 1998

All rights reserved; no part of this publication may be reproduced, stored in any retrieval system, or transmitted in any form or by any means, electronic, mechanical, photocopying, recording, or otherwise without either the prior written permission of the Publishers or a licence permitting restricted copying in the United Kingdom issued by the Copyright Licensing Agency Ltd, 90 Tottenham Court Road, London W1P 9HE.

First published 1998

British Library Cataloguing in Publication Data
a catalogue entry for this title is available from the British Library

ISBN 0-582-36856-1

Designed and set by Maria Pack and Brett Gilbert
Printed in Great Britain by Henry Ling Ltd, at the Dorset Press, Dorchester, Dorset

CONTENTS

PREFACE *ix*

CHAPTER 1   INTRODUCTION   *1*

CHAPTER 2   BALANCE OF AIR AND WATER FORCES   *3*
    2.1   Axes of Motion and Nomenclature   *3*
    2.2   Equilibrium States   *6*
    2.3   Typical Behaviour   *11*

CHAPTER 3   STATICAL STABILITY   *14*
    3.1   Generation of Righting Moment   *14*
    3.2   Righting Arm or GZ Curve   *15*
    3.3   Influence of Hull Shape   *16*
    3.4   Free Surface Effects   *22*
    3.5   Movable Ballast   *23*
    3.6   Longitudinal Stability   *25*
    3.7   Flooded Stability   *25*

CHAPTER 4   AERODYNAMICS OF SAILS   *27*
    4.1   Introduction   *27*
    4.2   Airflows around Thick and Thin Aerofoils   *28*
    4.3   Modern Aerofoil Design Methods   *31*
    4.4   The Relation of Pressure Distribution to Foil Geometry   *31*
    4.5   Three-dimensional Effects on Lifting Foils   *33*
    4.6   Identifying Induced Drag   *35*
    4.7   Identifying the Maximum Forward Drive Force   *36*
    4.8   Optimum Planform and Twist for a Single Sail   *37*
    4.9   The Effect of a Gap at the Sail Foot   *39*
    4.10 Three Dimensional Effects in Separated Flow Regions   *39*
    4.11 Sail Interactions   *40*
    4.12 The Effect of Heel on Sail Performance   *41*

       4.13 Reaching and Downwind Sails *42*
       4.14 Centres of Effort of a Sail Plan *43*
      References *45*
      Bibliography *45*

CHAPTER 5 **THE HYDRODYNAMICS OF HULL, KEEL AND RUDDER** *46*
       5.1 Introduction *46*
       5.2 Viscous Resistance *48*
       5.3 Wavemaking Resistance *52*
       5.4 Heeled Resistance *60*
       5.5 Induced Resistance *63*
       5.6 Hydrodynamic Sideforce *71*
      References *76*

CHAPTER 6 **DYNAMIC BEHAVIOUR OF SAILING YACHTS IN WAVES** *78*
       6.1 Introduction *78*
       6.2 Environmental Conditions *78*
       6.3 Motions in Waves *81*
       6.4 Added Resistance in Waves *89*
       6.5 Added Resistance in Following Waves and Surfing *99*
       6.6 The Steering Performance of the Yacht *101*
      References *108*

CHAPTER 7 **VELOCITY PREDICTION PROGRAMS** *109*
       7.1 Background *109*
       7.2 Aims of a VPP *110*
       7.3 Methodology *112*
       7.4 Hydrodynamic Force Model *116*
       7.5 Aerodynamic Force Model *127*
       7.6 Solution and Optimisation Routines *135*
       7.7 Presentation *139*
      References *143*

CHAPTER 8 **MATERIALS IN CONSTRUCTION** *145*
       8.1 Parameters Influencing Choice *145*
       8.2 Steel *150*
       8.3 Aluminium Alloys *151*
       8.4 Wood *155*
       8.5 Composites *158*
       8.6 Conclusions *160*
      References *161*

CHAPTER 9 **STRUCTURAL DESIGN OF HULL ELEMENTS** *163*
       9.1 The Aim of Structural Design *163*

Contents vii

9.2 What You Need *166*
9.3 Basic Structural Design Principles *168*
9.4 Loads *173*
9.5 Principles of Efficient Structures *178*
9.6 Safety Factors *182*
9.7 Deflection Criteria *185*
9.8 Hull Shell Dimensioning *186*
9.9 Dimensioning of Frames and Frame Systems *187*
9.10 Conclusions *188*
References *189*

CHAPTER 10 MAST AND RIGGING DESIGN *191*
10.1 Introduction *191*
10.2 Mast and Rigging *192*
10.3 Structural Aspects *194*
10.4 Mast Arrangement *196*
10.5 Design Methods *204*
10.6 Other Classification Society Rules *213*
10.7 Conclusions *213*
References *214*

CHAPTER 11 HULL DESIGN AND GEOMETRY DEFINITION *215*
11.1 Geometry of Hull Lines *215*
11.2 Computer Fairing *216*
11.3 Other Surface Definitions *227*
11.4 Approaches to Design *230*
11.5 Links to Other Design Software *232*
References *233*

CHAPTER 12 BACKGROUND TO COMPUTATIONAL FLUID DYNAMICS *235*
12.1 Introduction *235*
12.2 Potential Flow Methods *236*
12.3 Navier-Stokes *244*
12.4 Solution Techniques *247*
12.5 Grid Generation *248*
12.6 Visualisation and Validation *250*
12.7 Common Applications *252*
12.8 Closure *256*
References *257*

CHAPTER 13 BACKGROUND TO FINITE ELEMENT ANALYSIS *258*

13.1 The Role of FEA in Design *258*
13.2 FEA Theory – The Stiffness Method *261*
13.3 Factors Influencing the Use of FEA *272*
13.4 Conclusions *277*
References *277*
Bibliography *277*

CHAPTER 14 MODEL TESTING *278*

14.1 Introduction *278*
14.2 Facilities and Approaches *279*
14.3 Towing Tank Test Techniques *285*
14.4 Wind Tunnel Testing *298*
References *308*

CHAPTER 15 SAFETY ENGINEERING *310*

15.1 Introduction *310*
15.2 Basic Concepts *310*
15.3 Four Levels of Structural Reliability Methods *315*
15.4 Target Reliability Levels *323*
15.5 Some Typical Distributions *323*
15.6 Material Factors for Level I Design: Fibre Reinforced Composites *325*
15.7 Methods for Assessing Overall Safety *327*
15.8 Quantitative Risk Analysis and Formal Safety Assessment *327*
15.9 Application of Safety Engineering Principles to Sailing Yachts *331*
References *331*

INDEX *333*

PREFACE

Sailing craft form an expanding sector of the marine industry and events such as the America's Cup and the Volvo Ocean Race (previously the Whitbread Round-the-World Race) are receiving increased public interest. The leisure industry forms a substantial portion of the activities of many countries; water sports and sailing boats are of principal interest here. The science and technology associated with the design, construction and operation of sailing yachts, along with fabrication and construction materials technology, are all advancing at a rapid rate; and new design tools based on computational techniques are emerging. It is important for students of the subject and industrial designers to be aware both of the physical mechanisms by which sailing craft are governed and the science and technology employed in their design.

This book and its companion volume – *Sailing Yacht Design: Practice* which deals with practical aspects – are intended to provide a sound theoretical base for the design, manufacture and operation of sailing craft. They represent a first step in further study of this interesting and rewarding subject.

Sailing Yacht Design: Theory contains the fundamental theory concerning the hull and sail static force equilibria, shape definition and associated stability. There is a discussion on the aerodynamics and hydrodynamics of the sail, hull, keel and rudder and their influence on the seakeeping abilities of the yacht. There is a brief section on the different materials used in hull construction and chapters on structural design techniques and mast/rigging design. There is then coverage of different computational and experimental techniques that help in good design. Finally, safety as a topic of growing importance provides an overall backdrop to design.

Sailing Yacht Design: Practice contains a series of chapters by different designers on their experiences of translating the scientific principles into reality. There is coverage of the practical design of hulls, appendages such as keels and sails. An important feature is the practical design of the structure of a high-performance hull made from advanced composites. Production techniques and boatyard facilities are covered separately. Finally, there are some chapters on handicap rules, statutory and regulatory constraints, and safety considerations from a practical perspective.

the authors of the various chapters in the two books are all internationally renowned authorities. They are professionally engaged in the field of sailing yacht design, construction, design consultancy, classification societies, yachting associations, materials supply, research establishments and universities. Without the tremendous effort put in by the authors, and their cooperation in meeting deadlines, these books would not have been possible. The editors wish to thank the authors for all their help and assistance.

the material in the two books was compiled for a WEGEMT School held in the University of Southampton in September 1998. WEGEMT is an association of European universities in Marine Technology, which exists to promote continuing education in this broad field, to encourage staff and student exchanges and to foster common research interests.

Participants at such schools have generally been drawn from the ranks of professional engineers and naval architects in shipyards, boatyards, consultancies, etc. A large proportion have also been postgraduate students and staff wishing to obtain an overview of a particular topic as a basis of research.

the Southampton WEGEMT School was organised with the help of an international steering committee whose members were:

Mr G. Belgrano	*SP Raceboat Group*
Mr A.R. Claughton	*Wolfson Unit for Marine Technology and Industrial Aerodynamics*
Dr G. Dijkstra	*Ocean Sailing Development BV*
Mr E. Dubois	*Dubois Naval Architects*
Mr G. Holm	*VTT*
Dr J.A. Keuning	*Delft University of Technology*
Mr P. Morton	*Farr International*
Dr R.A. Shenoi	*University of Southampton*
Dr J.F. Wellicome	*University of Southampton*

the committee approved the course content and helped select the course lecturers whose notes form the material of these two books. The editors are grateful to the members for their advise and guidance.

the School was supported in part by funds from the European Union under the TMR programme. We are indebted to Mr J.A.T. Grant, Secretary General of WEGEMT, for his help in obtaining the TMR funding and publicising the School.

the encouragement, support and assistance given by Professor W.G. Price and colleagues in the Department of Ship Science and the Wolfson Unit for Marine Technology and Industrial Aerodynamics has been most generous and invaluable. We are grateful to them, and want to acknowledge the particular assistance of Mr Jason Smithwick. Finally, we wish to extend our thanks to Dr Maria Pack for her expertise and professionalism in preparing the manuscripts, undertaking the word processing and for patiently coping with the numerous edits, changes and amendments involved in preparing the camera-ready copy for the two books.

A.R. Claughton, J.F. Wellicome, R.A. Shenoi
Southampton

CHAPTER 1
INTRODUCTION

A. Claughton*, A. Shenoi[†], J. Wellicome
*Wolfson Unit for Marine Technology and Industrial Aerodynamics,
University of Southampton, [†]University of Southampton.

Sailing yachts are uniquely complex vehicles. They operate at the interface of two fluids, air and water, deriving propulsion from the former and support from the latter. Both these media are subject to atmospheric effects, the wind can blow a gale or not at all, the sea can be mirror smooth or violently rough. Not surprisingly, in this complex environment, many aspects of yacht design resist perfect mathematical analysis, and consequently the design of a sailing yacht also straddles two media; namely art and science. Successful yachts are often artefacts of real beauty, but this is a reflection of the fact that correct analytical design can be married to aesthetically pleasing physical form, rather than the old maxim that 'what looks right is right'.

Sailing yachts offer a wide scope for scientific investigation, presenting problems of the highest complexity, not only the fluid mechanics aspects, but also the solid mechanics considerations and the complex interactions between form and weight which determine the yacht's performance and behaviour.

The aim of this book and its companion volume – *Sailing Yacht Design: Practice* – is two-fold:

1. To provide the yacht design community with a comprehensive text on both the theoretical and practical aspects of sailing yacht design
2. To provide students and researchers in related fields with a starting point for their research

This book deals mainly with scientific and theoretical considerations underpinning the art of sailing yacht design. The emphasis throughout has been on the mechanics aspects. The contents of this book can be divided into four broad categories – fluid mechanics, solid mechanics, computational/experimental mechanics and safety engineering.

The first category of subjects covered here deals with fluid mechanics. The book begins with an examination of the force system acting on the yacht, both in terms of water- and air-based loads. This is then used to evaluate the statical stability of the hull forms in both intact and damaged conditions. Next, more in-depth coverage is given to the aerodynamics of sails. Included in this is a treatment of aerofoil design methods, planform and twist for single sails, sail interactions and reaching and downwind sails. This is followed by an equivalent treatment of the hydrodynamics aspects involving the hull, keel and rudder interaction; the essential thrust of the coverage is to estimate various drag or resistance and sideforce components. An understanding of the dynamic behaviour of a yacht is particularly important for good design; this follows naturally from the hydrodynamic study. Finally the various aspects of fluid dynamic design are drawn together in a chapter on performance prediction methods employed in velocity prediction programs (VPPs). There is extensive coverage on the methodology of a VPP, the hydrodynamic and aerodynamic force models and the optimisation routines.

The second category of subjects deals with solid mechanics aspects. This is prefaced by an examination of the materials that could be used in yacht construction, including steel, aluminium, composites and wood. Then there is a detailed treatment of the structural design of hull elements, beginning with the evaluation of structural loads on the hull girder owing to hydrodynamic elements followed by design principles and dimensioning of the plating and framing components. Next a similar approach is followed for the mast and rigging design, which is the main based on aerodynamic loading through the sails. There is coverage of different mast arrangements and of typical design methods used by classification societies.

The third category deals with computational and experimental techniques to aid the design process. This is naturally split into two sub-categories, the first of these dealing with the computational aspects. Three facets under this heading relate to the definition of the hull form, fluid flow modelling and structural analysis modelling. A good hydrodynamic hull form requires a precise definition of the hull shape; the mathematical bases for form definition are covered in some depth as are some of the practical issues related to software use. Next, there is a broad outline of the computational fluid dynamics (CFD) principles, including both potential flow and Navier-Stokes methods. There is some treatment of grid generation and visualisation/validation techniques in this context. The equivalent treatment on the structural front requires an introduction to finite element methods (FEM); an outline is given of the role of FEM in *design*, some essential fundamentals are introduced and factors affecting the use of FEM are discussed. This category concludes by examining the experimental techniques available for validating some of the computational predictions; there is coverage of both hydrodynamic (towing tank) and aerodynamic (wind tunnel) testing.

The final category has been specially introduced with a view to reinforce the growing importance of formal safety methods in design. This trend is becoming the norm in other naval architectural and engineering design situations. It is essential that yacht designers also consider the role of statistical tools to improve the quality of their product.

CHAPTER 2

BALANCE OF AIR AND WATER FORCES

A. Claughton
Wolfson Unit for Marine Technology and Industrial Aerodynamics,
University of Southampton

2.1 AXES OF MOTION AND NOMENCLATURE

Mechanism of sailing

The sailing yacht operates at the interface between two fluids, the immersed part of the hull moves through water along the yacht's 'track', the above water part of the hull, the mast and sails move through the air, experiencing an incident wind field that is the vector sum of the natural or true wind V_T and the wind arising from the yacht's motion through the air. The wind 'felt' by the sails is termed the apparent wind (V_A).

Figure 2.1 shows a yacht sailing at a steady speed in calm water, and in this quasi static equilibrium condition the net forces and moments acting on the vessel are zero. The forces acting on the above water part of the yacht (FA_{TOT}), propel the underwater part, which produce an equal and opposite force (FH_{TOT}). By adjustment of the sails and rudder the yacht can be made to hold a steady course. The waterplane, although not strictly a plane once deformed by the hull wave system, is conventionally treated as the divide between the aerodynamic and the hydrodynamic forces. The behaviour of the yacht can be viewed as a balancing of the forces in these two domains.

4 Sailing Yacht Design: Theory

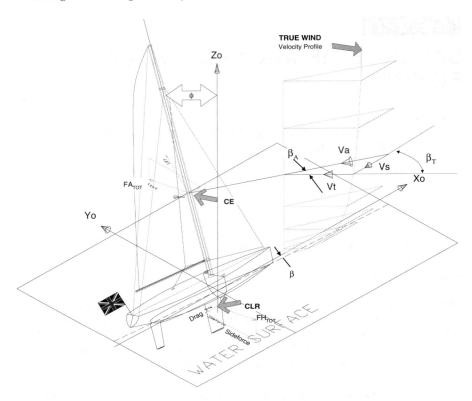

Figure 2.1 *Forces acting on a sailing yacht.*

Axis system

To study these effects in more detail appropriate axes systems for the aerodynamic and hydrodynamic parts of the yacht must be defined.

Space axes

Hydrodynamic characteristics are usually presented on 'track axes' that is a right-hand orthogonal axis system that is in the vertical and horizontal (water surface) planes with the Xo axis aligned with the yacht's direction of motion, Yo is positive to port, and Zo positive upwards. The corresponding moments are roll moment M_{Xo}, pitch moment M_{Yo}, and yaw moment M_{Zo}.

The important forces and velocities acting in this plane are:

- V_S (speed), velocity along Xo
- FH_{Xo} (drag), the component of hydrodynamic force acting along Xo
- FH_{Yo} (sideforce), the component of hydrodynamic force acting along Yo

Figure 2.1 shows the yacht sailing on starboard tack, i.e. the apparent wind blows over the starboard side of the yacht. The yacht heels to port under the influence of the negative Mx created by the sails, the hull produces a negative FH_{Yo} to starboard to counteract this by adopting a yaw angle (β). The hull centreline is now at an angle to its direction of motion and we must establish a further axis system that remains fixed to the yacht, the body axis system.

Body axes

Structurally based characteristics are generally considered in a co-ordinate system that moves with the body of the yacht, it is another right-hand orthogonal system based on the centre plane of the yacht, aligned with hull centreline and mast. X forward along the yacht centre line, Y normal to the centreline of the yacht in the plane of the deck, and Z vertical, in the plane of the mast.

Wind axes

Aerodynamic characteristics of sail plans are often expressed as in terms of lift (Cl) and drag (Cd) coefficients. The aerodynamic lift and drag force vectors are normally considered to lie in the Xo, Yo plane but are aligned with and at right angles to the apparent wind.

True wind

The true, or natural, wind is modified by having moved across the water (and in the case of inland sailors over the land) surface. This creates a boundary layer where under normal circumstances the wind blows more strongly as height above the sea increases. Also in some cases the wind direction may vary with height above the ground, although this rarely happens to a discernible extent over the height of a yacht mast. The velocity and direction of the true wind are defined as:

V_T = true wind speed (m/s) measured at a known height above the water surface

β_T = true wind angle measured between the yacht's track and the V_T vector

Depending on atmospheric conditions this wind gradient may be expressed in the following equation:

$$V_T(z) = V_T(z_{ref}) \cdot (z/z_{ref})^\alpha \qquad [2.1]$$

Z = height above Xo Yo plane
Z_{ref} = reference height for V_T measurements

The exponent α can vary between 1/7 and 1/14 depending on wind and atmospheric conditions.

The velocity at any height is obviously not steady and the above equation predicts mean velocities, the natural wind also contains broad spectrum turbulence which can affect the aerodynamic behaviour of the aerodynamic components. The most commonly used reference height for wind speed measurements is 10 m above the water surface.

Apparent wind

The yacht sails in a wind field that is the vector sum of the true wind and the apparent wind created by the passage of the yacht through the air. Figure 2.1 shows how the apparent wind speed (V_A) and the apparent wind angle (β_A) change with height above the water surface. Equations 2.2 and 2.3 show how V_A and β_A are calculated.

$$\beta_A = \tan^{-1}[V_T \sin \beta_T \cos \phi /(V_T \cos \beta_T + V_S)] \qquad [2.2]$$

$$V_A = \sqrt{[(V_T \sin \beta_T \cos \phi)^2 + (V_T \cos \beta_A + V_S)^2]} \qquad [2.3]$$

The heel angle is included in these equations because the apparent wind field is often considered to move with the centreplane of the yacht when calculating aerodynamic behaviour.

Reference height
Once the yacht's motion and the true wind have combined to yield an apparent wind velocity the appropriate reference height must move into the boat frame of reference. Apparent wind velocities are generally calculated at the sail plan centre of effort (CE), for the purposes of calculating force coefficients.

2.2 EQUILIBRIUM STATES

The aerodynamic and hydrodynamic forces and moments must be in equilibrium for the yacht to maintain a steady course. In the following sections these equilibria are defined by simple equations, and also the physical parameters that influence the terms are shown, based on the nomenclature of Figure 2.1.

FA_{TOT} is assumed to act through the CE, and FH_{TOT} through the centre of lateral resistance (CLR). To maintain a steady course the components of FH_{TOT} and FA_{TOT} in the horizontal plane must be equal and opposite, as shown in Figure 2.2 which is a view along the Zo axis of Figure 2.1.

Balance of Air and Water Forces 7

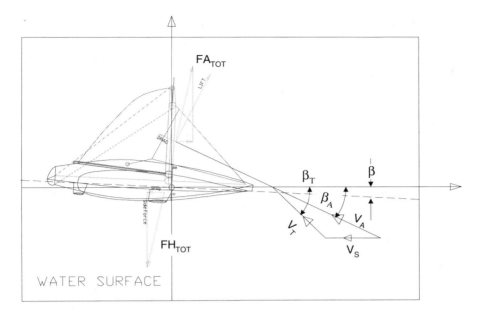

Figure 2.2 *Components of forces acting on a sailing yacht in the plane of the water surface.*

Aerodynamic and hydrodynamic forces

The aerodynamic lift (L) and drag (D) forces are assumed to act normal to the centre plane of the hull and mast and may be resolved into the space axis system by the following equations:

$$FA_{Xo} = L \sin \beta_A - D \cos \beta_A \qquad [2.4]$$

$$FA_{Yo} = (L \cos \beta_A + D \sin \beta_A) \cos \phi \qquad [2.5]$$

When the yacht is sailing in a steady condition the transverse component of the aerodynamic force (FA_{Yo}) is opposed by the sideforce (FH_{Yo}), which is generated by the hull adopting an angle of leeway (β). The leeway angle will increase until the hull and keel and rudder produce sufficient sideforce to balance the aerodynamic force. Besides the component that heels the yacht the sails also produce, if correctly trimmed, a force (FA_{Xo}) that can move the yacht forward, and under the influence of this force the yacht's speed will increase until the hydrodynamic drag (FH_{Xo}) is the same as the aerodynamic propelling force.

Driving force and resistance

This is the fundamental equilibrium that determines the yacht's speed, in simple terms if resistance can be reduced or driving force increased then the yacht will sail faster. It must be remembered however that an increase in driving force is usually accompanied by an increase in heeling force, which will cause the vessel to heel more.

Resolving forces along the yacht's track: the driving force (FA_{Xo}) is equal and opposite to the Resistance (FH_{Xo}) which is dependent on V_S, ϕ and β.

Sideforce and heeling force

This equilibrium determines the leeway angle that the hull must adopt to resist the heeling force of the sails. Resolving the forces along the Yo axis: FA_{Yo} is equal and opposite to the sideforce (FH_{Yo}). The equilibrium leeway (β) is a function of Sideforce, V_S and ϕ.

Vertical force

In the vertical axis the components of the aerodynamic and hydrodynamic forces are joined by the gravitational forces, the weight of the yacht (W) and the buoyancy force (B). Any difference between the vertical components from the aerodynamic and hydrodynamic forces is accommodated by a change in draught of the hull which modifies the buoyancy force.

Roll moment (Mx)

The roll moment balance is the most crucial equilibrium for the yacht. The mechanism of hydrostatic stability is described in detail in Chapter 3, but it is appropriate to examine the effects at this point.

Roll equilibrium is most easily considered in the YZ plane of the yacht. The couple between the aerodynamic and hydrodynamic forces heels the yacht, as the yacht heels a counteracting righting moment (RM) is generated by the separation of the vertical gravity forces, weight (W) and buoyancy (B).

Water flow over the hull surface and hull generated waves create a small modification to the righting moment, MH_X, which is a function of speed, heel and leeway.

Sailing sideforce

The roll moment equilibrium shown in Figure 2.3 is such that for a given yacht the hydrodynamic sideforce (FH_{Yo}) at any heel angle can be shown to depend simply on the yacht's hydrostatic characteristics and the position of CE and CLR.

The total aerodynamic force is usually assumed to act normal to the mast. The component of this force in the YZ plane is termed the heeling force (FH).

$$FH = SF/cos\phi \qquad [2.6]$$

$FH \cdot a$ = righting moment (RM) = $W \cdot GZ$ [2.7]

Therefore at any given heel angle, ϕ:

$$FH_\phi = \frac{RM_\phi}{a}$$ [2.8]

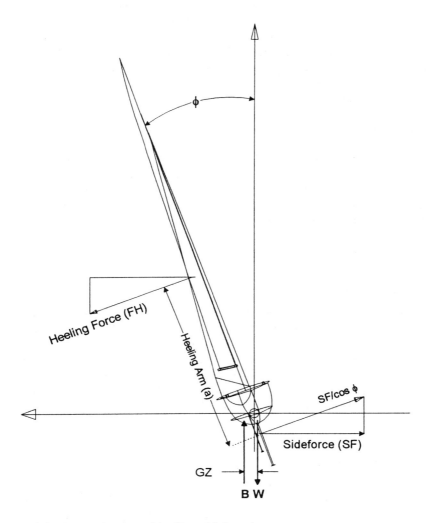

Figure 2.3 *Determination of 'sailing sideforce'.*

From equation 2.8 it can be seen that the equilibrium heeling force (FH) at heel angle ϕ is a function of the righting moment and the separation of the CE and CLR, this means that for a given sailplan geometry and hull VCG, at any heel angle the equilibrium heeling force is **fixed**, apart from the second order changes caused by hull wave pattern effects.

In simple terms, if the hull is made more stable, then the equilibrium heeling force at a given heel angle increases and if the sailplan is made taller, and the CLR–CE separation is increased then the equilibrium heeling force is reduced.

This relationship between the hull stability and sailing sideforce is crucial to how a yacht sails and what factors affect its performance. At any heel angle the heeling force is determined by the roll equilibrium conditions; the hull then adopts a leeway angle so that the hydrodynamic force is equal and opposite to the aerodynamic force. The leeway angle will be determined by the speed of the vessel and the required sideforce (FH_{Y_0}). This equilibrium hydrodynamic sideforce is termed the 'sailing sideforce'.

Righting moment

One of the features of sailing yachts is that the righting moment is not completely determined by the VCG position of the hull and rig structure because the crew and gear can move to windward, and in some cases water ballast is added on the windward side. This shifts the centre of gravity and thereby increases GZ and hence righting moment. Thus the effect of the crew 'hiking' is to increase the yacht's sailing sideforce at a given heel angle, although of course, once the yacht's stability has been augmented the apparent wind speed required to reach that heel angle will increase.

Yaw moment (M_{Z0})

Figure 2.2 shows that the components of FA_{TOT} and FH_{TOT} in the plane of the water surface must have coincident lines of action. Examination of Figure 2.2 shows that if CE remains in a fixed longitudinal position as the yacht heels more the CLR must move aft to retain coincident lines of action. Control of the position of CLR is provided by the rudder, increasing the rudder angle of attack shifts the CLR aft. Some control over the position of CE can be exerted by adjusting sail trim, for example reducing the mainsail angle of attack by easing the main sheet shifts the CE forward.

Pitch moment (M_{Y0})

The aerodynamic driving force and the hydrodynamic drag produce a significant bow down trimming moment. At low speeds this is countered by the hull trimming to produce an opposing couple between the gravity and buoyancy forces. At higher speeds the buoyancy forces are augmented by hydrodynamic lift from the forward sections.

Sail force trimming moment

The bow down trimming moment exerted by the couple between aero and hydrodynamic forces is sometimes referred to as the sail trimming moment, and this has an important part to play in the force balance that must be implemented during sailing yacht model tests.

2.3 TYPICAL BEHAVIOUR

The sailing speed and interaction of the aerodynamic and hydrodynamic forces are strongly influenced by both the strength of the wind and the true wind angle.

Figure 2.4 shows how the yacht's sailing speed may be presented graphically as a polar diagram. At each true wind speed the yacht's speed is plotted as a displacement along the true wind angle line. Thus the wind may be thought of as blowing from the top of the diagram and the yacht's speed at each true wind angle radiate from the origin like spokes of a wheel. Polar curves for three true wind speeds are shown in Figure 2.4.

Figure 2.1 shows the yacht sailing 'close hauled', that is sailing with a true wind angle of approximately 45°. This condition is characterised by the aerodynamic heeling force being several times greater than the driving force. Thus the yacht adopts a relatively high heel angle and the sailing speed is relatively low. This point is shown on Figure 2.4 (V_S = 8.2 knots). In this situation the helmsman of the yacht is attempting to make ground directly to windward, and is therefore trying to maximise his velocity made good (V_{MG}) to windward. From simple geometry:

$$V_{MG} = V_S \cos \beta_T \qquad [2.9]$$

If he sails too close to the wind, that is at too low a true wind angle, while his boat speed is directed more nearly in the direction he wants to go, the available aerodynamic force is reduced and the yacht's speed falls. If he sails too far off the wind, that is at too high a true wind angle, then while more aerodynamic force is available and V_S is increased, the boat speed vector is directed too far from the true wind direction. The skill of the crew is to adjust the sails and the course to occupy the point on the polar curve that has the maximum V_{MG}. In a true wind of 16 knots the maximum V_{MG} is 6.1 knots.

12 *Sailing Yacht Design: Theory*

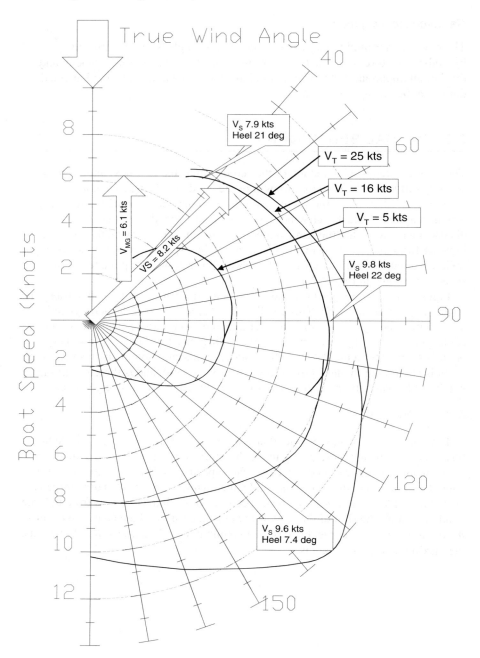

Figure 2.4 *Typical 3 m LOA sailing yacht polar diagram.*

As the yacht sails at greater true wind angles then more aerodynamic driving force becomes available. Although heeling force is still high, the driving force is of a similar magnitude, and V_S increases to 9.8 knots.

Once the yacht has borne away from the wind to a true wind angle of 135° the boat's speed is reducing the apparent wind and thus sails of greater area may be set to increase the available aerodynamic driving force. Also the magnitude of the heeling force compared to the driving force is reduced, and so the control of excessive heeling moment is no longer a major concern to the crew. This condition, shown on the polar curve in Figure 2.4.

Chapter 7 describes in more detail the prediction of sailing speed, the polar curves in this chapter have been included to provide a background to the discussion of the aerodynamic, hydrostatic and hydrodynamic forces in Chapters 3–6.

CHAPTER 3
STATICAL STABILITY

B. Deakin
Wolfson Unit for Marine Technology and Industrial Aerodynamics,
University of Southampton

3.1 GENERATION OF RIGHTING MOMENT

Yachts resist heeling by the generation of a righting moment, which in turn is the result of movement of the centre of buoyancy relative to the centre of gravity. The moment is the product of the yacht's displacement and the lateral separation of these centres, which is known as the righting arm. Figure 3.1 shows the conventional nomenclature associated with the presentation of stability data.

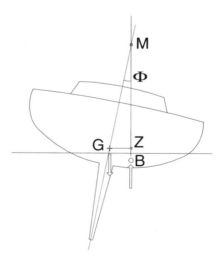

Figure 3.1 *Nomenclature associated with the generation of righting arm.*

As the yacht heels, the centre of buoyancy, B, moves to one side of the centreline. The buoyancy force acts vertically upwards, and the intersection of its line of action with the yacht's centreline is known as the metacentre, M. For small angles of heel, the movement of the centre of buoyancy tends to be proportional to the angle, and so the righting arm, GZ, and righting moment are also proportional to the angle. At small angles the righting arm is equal to GM sin ϕ, and so GM provides a simple measure of the initial stability, or 'stiffness' of the yacht.

3.2 RIGHTING ARM OR GZ CURVE

With increasing heel angle, the lateral separation of the centres of buoyancy and gravity reaches a maximum, and subsequently they move back into the same vertical plane. GZ correspondingly increases to a maximum value, and then reduces to zero. The angle at which GZ equals 0 defines the limit of positive stability, normally referred to as the range of stability. See Figure 3.2. If released at an angle less than ϕ the yacht will return to upright, but if released at a greater angle it will continue to heel, and will capsize.

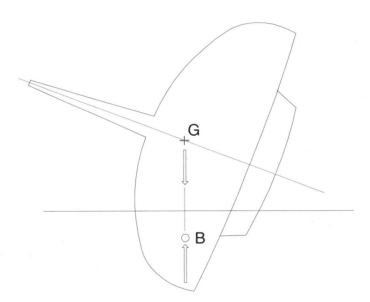

Figure 3.2 *A yacht heeled to the limit of its range of positive stability.*

It is possible for the yacht to have a range of stability of 180°, and thus be completely self righting, but it is usual to have a range of angles over which the stability is negative. If the yacht is released at an angle within this range it will settle upside down.

Figure 3.3 illustrates the conventional graphical presentation of stability, that is the GZ curve, or variation of GZ with angle of heel, and shows the characteristics described above. When comparing yachts, or assessing them against some standard, it is normal to refer only to the righting arm. Although GM and GZ are not non-dimensional parameters, they generally tend to be independent of vessel size.

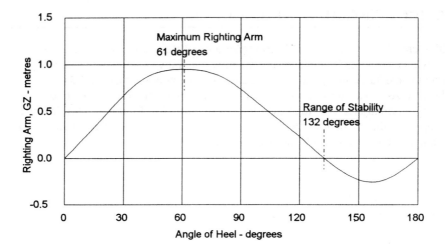

Figure 3.3 *A typical righting arm, or GZ curve.*

3.3 INFLUENCE OF HULL SHAPE

All aspects of the hull shape affect the stability to some extent, but one should be cautious about examining the effects of individual parameters in isolation. A yacht design is developed to fulfil a set of requirements, and this will guide the choice of hull form and outfit. The decision to change a particular parameter will force or enable changes in other parameters, or in the weight distribution. The effect on stability of changing a parameter therefore may not be obvious.

Beam and centre of gravity

The stability characteristics are usually dominated by the beam and the height of the centre of gravity. A lower centre of gravity will always provide increased stability at all angles of heel and, in general, every designer and builder should strive to maintain as low a centre of gravity as possible in order to maximise sailing performance and safety.

The value of BM is equal to the transverse second moment of area of the waterplane divided by the displacement, and therefore may be considered to be proportional to the square of the yacht's beam. By increasing the beam, and hence BM, the designer will be able to reduce the weight of ballast while retaining the same GM and ability to carry sail. Thus an increase in beam is often associated with a reduction in displacement and an increase in the centre of gravity height.

The relationship between beam, centre of gravity and stability is illustrated in Figure 3.4, in which examples of three characteristic types of yacht are compared at four angles of heel. On the left is a traditional cruising yacht form, characterised by heavy displacement, narrow beam, and a low centre of gravity achieved by locating ballast at the bottom of a long keel. In the centre is a modern cruiser/racer, having lighter displacement, greater beam, and a higher centre of gravity with ballast distributed throughout a fin keel. The third type illustrated, taking wide beam to the extreme, is a cruising catamaran. Having no ballast, this type of yacht has a light displacement and a relatively high centre of gravity.

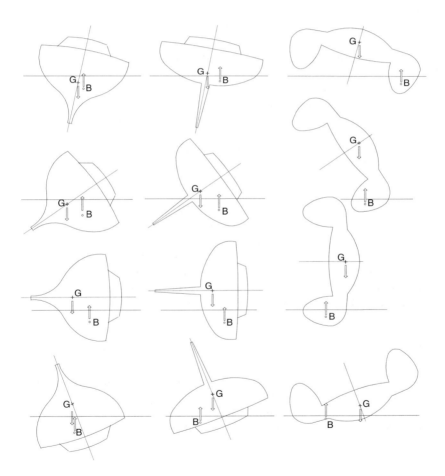

Figure 3.4 *The effect of hull form and centre of gravity on stability.*

The yachts are illustrated at angles of heel of 10, 65, 90 and 160°, and the centres of gravity and buoyancy are indicated for comparison. Bearing in mind that GZ represents the lateral separation of these centres, these sketches will help to explain the differences between the GZ curves for these three types of yacht which are presented in Figure 3.5.

The traditional cruiser has relatively low initial stability because its waterline beam is small. The stability continues to increase with heel angle up to 90°, and the range of stability is in excess of 160°.

The wider modern form has greater initial stability, but this reaches a maximum at about 60°. The higher centre of gravity results in a comparatively lower righting arm at 90°, and, combined with wide beam and high flotation, results in a lower range of stability, in this case 132°. This form has significant stability in the inverted state, enabling it to remain inverted if capsized.

The catamaran form is renowned for its high stability, but this is restricted to small angles. The extreme beam provides sufficient stability to counter the wind heeling moments without the need for ballast. The righting arm increases rapidly with heel angle as more of the displacement is supported by the leeward hull, and the centre of buoyancy moves to that side. It reaches a maximum when the windward hull emerges from the water, as shown in the first diagram, at less than 10°. The righting arm subsequently reduces roughly as the cosine of the heel angle, but with a range of stability typically less than 90°. The range is governed by the height of the centre of gravity which, in the absence of ballast, is relatively high. The stability when inverted is similar to that when upright, being dominated by the wide beam, and a capsized catamaran is unlikely to be righted without external assistance.

These stability curves provide a good comparison of the hull types in terms of the variation of righting arm over the heel angle range, but it must always be borne in mind that the displacement needs to be included if the resistance to wind heeling, or other external moments, is to be quantified. To put this into perspective, compare the righting arm curves in Figure 3.5 with the righting moment curves presented in Figure 3.6. These latter curves have been drawn assuming displacements for the three yachts of 5.5, 4 and 3 tonnes respectively. Whilst the shapes of the curves remain the same, their magnitudes have been brought closer together.

If all three were to carry the same rig, they would be subjected to the same heeling moment, and an example is shown on Figure 3.6, where a common heeling moment curve has been superimposed on the righting moment curves. A yacht will be in equilibrium in terms of its heel angle when the righting moment is equal to the heeling moment, and in the example shown the traditional yacht would heel to 30°, while the modern form would heel to 25°, and the catamaran to less than 4°.

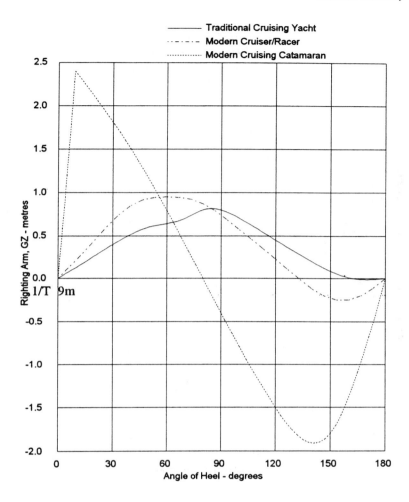

Figure 3.5 *Righting arm curves for the yachts illustrated in Figure 3.4.*

Freeboard

The linear nature of the GZ curve normally is restricted to a range of angles within which there is no substantial change in the shape of the waterplane. Immersion of the deck edge, or emersion of the turn of bilge, limits this range and so the freeboard has a significant effect on the stability at angles beyond deck edge immersion.

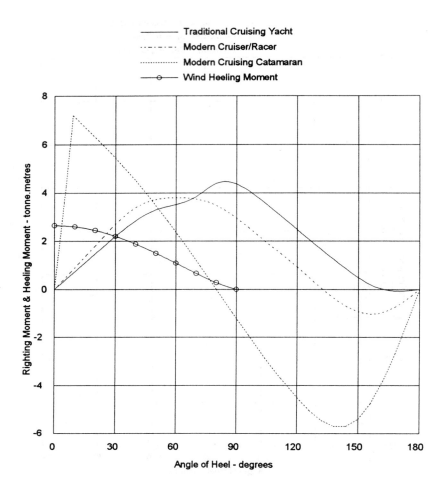

Figure 3.6 *Righting moment curves for the yachts illustrated in Figure 3.4.*

Two GZ curves are presented in Figure 3.7 to illustrate the magnitude of the effects of freeboard. A yacht designed with a freeboard of 1.0 m has been modified by increasing the height of the deck, while maintaining the same levels for the top of the coachroof and the cockpit sole. To isolate the effect of the hull geometry it has been assumed that the centre of gravity remains unchanged, although of course this would be unlikely. The freeboard increase results in an increase in the angle of deck edge immersion, and in the maximum GZ achieved.

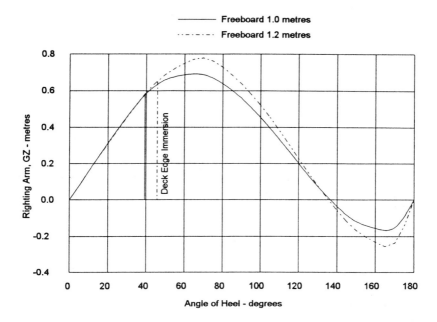

Figure 3.7 *The effects of freeboard.*

Coachroof and cockpit volumes

A large coachroof increases the stability at angles around 90° by moving the centre of buoyancy further from the centre of gravity. When inverted, the coachroof has a profound effect because, if large, the coachroof volume may be sufficient to support the yacht, and its relatively low beam will provide little inverted stability. If, on the other hand, there is little or no coachroof, the yacht will float on its deck when inverted and will be difficult to right.

Because, in the example illustrated in Figure 3.7, the coachroof top remained at the same level when the deck was raised, the volume of the coachroof above the deck was reduced and the stability when inverted was increased, to the detriment of safety.

The cockpit of a modern yacht has little effect on the stability because it remains above the waterline at most heel angles. Coamings with a high volume may have a beneficial effect when inverted, providing additional buoyancy at a favourable level.

Figure 3.8 shows the combined effects of a typical coachroof, cockpit and coaming arrangement on the stability.

22 *Sailing Yacht Design: Theory*

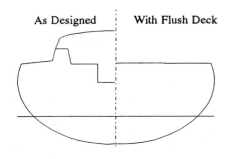

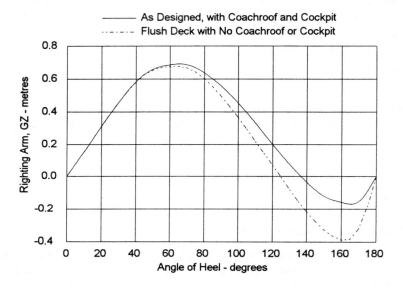

Figure 3.8 *The effects of the coachroof and cockpit.*

Hull flare

Figure 3.9 illustrates the effect of a 10% increase in the ratio of the beam at the deck to the waterline beam. Comparison with Figure 3.7 shows that increasing the flare has a similar effect to an increase in freeboard, increasing the stability at large angles, and making the yacht more stable when inverted.

3.4 FREE SURFACE EFFECTS

Partially filled tanks allow movement of liquid in response to an angle of heel, and this is often referred to as a free surface effect. It is normally taken account of in stability calculations by an adjustment of the VCG. This is an approximation, but is sufficiently accurate for the size of tanks found on yachts.

The adjustment is calculated as:

$$VCG\,fluid = VCG\,solid + \frac{Transverse\ 2^{nd}\ moment\ of\ area\ of\ liquid\ surface\ \times\ Liquid\ density}{Displacement} \qquad [3.1]$$

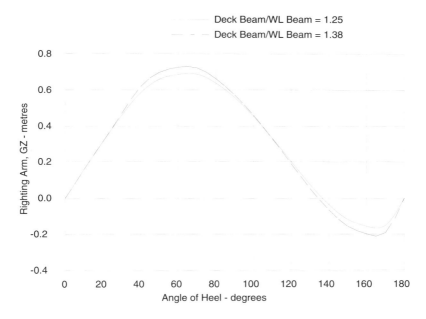

Figure 3.9 *The effects of hull flare.*

Note that reduction in the width of a tank by adding a central bulkhead will reduce the free surface effect by a factor of 4. It is common to install baffles within tanks to minimise slopping of the liquid, but these do not affect the movement of liquid in response to heel angles which remain relatively constant, as is the case on a heeled sailing yacht. Similarly, wing tanks which are cross connected, even by a small pipe, will be subject to transfer of liquid while the yacht sails heeled on a given tack, and may have a significant effect on the heel angle.

Wide tanks and cross coupling of tanks therefore should be avoided in the interests of performance, comfort and safety.

3.5 MOVABLE BALLAST

In order to improve sailing performance by a reduction in the heel angle, ballast is frequently moved, or added, to the windward side of the yacht. Methods of achieving this include moving the crew, swinging the keel, or pumping water ballast into a wing tank. Whilst the righting moments which result may be calculated readily for normal sailing heel angles, the effect that these actions have on the stability curve at larger angles will depend on the details of the system.

The crew are likely to change their position at extreme heel angles, and if the yacht capsizes they may, for example, be in the sea remote from the yacht, hanging onto some part of its rig or structure, or sitting on the inside of the coachroof. It is difficult to predict their effect on the ability of the yacht to right itself, and it is suggested that the crew should not be included when assessing the range of stability.

A swinging keel probably will incorporate some mechanism to ensure that its position remains fixed at all angles of heel, and the stability at large angles may be calculated by taking into account the offset centre of gravity of the yacht, if the computer software has this facility,

Pumped water ballast may remain in the tank at all angles of heel, even following a capsize, but this will depend upon the tank construction and venting arrangements. If such is the case its effect on the large angle stability may be assessed as for the swinging keel, bearing in mind that the displacement will also be altered by the additional ballast.

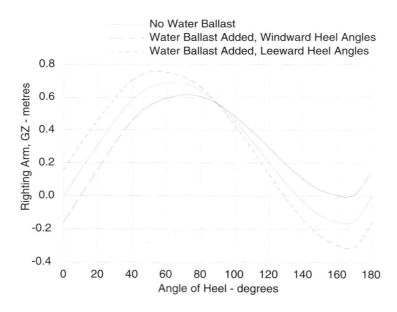

Figure 3.10 *The effects of water ballast in a wing tank.*

Figure 3.10 shows the effect of water ballast in a wing tank on the stability through 360°, assuming that the VCG and LCG remain unaffected by the addition of ballast. The ballast is sufficient to reduce sailing heel angles by 10°, and the yacht would heel to 10° in the absence of any external heeling moment. When considering heel angles to leeward, the ballast increases the maximum righting arm but reduces the angle at which the yacht will capsize. It does not reduce the overall positive range of stability however, since the GZ curve is positive from −10–129°, compared with 0–137° without the ballast. Furthermore, when heeled to windward, the yacht is shown to be virtually self-righting, and the only angles at which the yacht could remain inverted are therefore between 160–170° to windward. These values are highly dependent on the details of the yacht and its ballast arrangement, but they serve to counter the common misconception that offset ballast will jeopardise the safety of the yacht in the event of a capsize.

3.6 LONGITUDINAL STABILITY

The longitudinal stability of a conventional monohull yacht is an order of magnitude greater than the transverse stability, with longitudinal GM values typically in the range 8–15 times the transverse GM. The parameter therefore is of little concern in terms of safety unless the subject is a multihull.

Sailing multihulls are designed to a wide range of beam/length ratios, from 0.3 to 1, with the highest ratios generally applying to trimarans and high-performance racing catamarans. The longitudinal GM may be substantially less than the transverse GM and, although this does not necessarily result in a lower maximum righting moment in pitch, it does make multihulls vulnerable to deck immersion at the bow, which may in turn lead to a longitudinal capsize, or 'pitchpole'.

3.7 FLOODED STABILITY

Types of flooding and their effect on stability

A yacht may take on water as a result of minor leaks, downflooding through access or ventilation openings, or major damage below the waterline.

An alert crew will minimise the accumulation of water from minor leaks by pumping, and their effect on stability is unlikely to be significant. The bilge is sometimes considered as a tank, with its typical content and free surface taken account of in the loading condition of the yacht.

A downflooding event may be the result of a wave washing over the deck, or the yacht being heeled to a large angle. The quantity of water will be considerable if the opening is large, such as a companionway, or if the yacht is pinned down to a large angle for a prolonged period. In an extreme case such flooding, or subsequent wave action, might result in loss of the yacht, or the internal waterline being above the external waterline if the flooded space is a watertight compartment of limited size.

If the hull is damaged below the waterline, and leakage exceeds the capacity of the pumps, flotation of the yacht will rely on the maintenance of sufficient intact buoyancy.

Provision of adequate buoyancy when flooded may be achieved through intact buoyant structure or by adequate subdivision of the hull into watertight compartments. The former may be possible on a small craft, particularly with the aid of composite sandwich hull structures, but subdivision, or a combination of the two, is more likely to be practical on large ballasted yachts.

To facilitate the calculation, flooded and damage stability traditionally were approached by adding the appropriate weight and centre of gravity of floodwater to a vessel's loading condition, and applying a free surface correction. While this might be acceptable for the investigation of minor flooding, it does not offer sufficient accuracy for reliable damage assessment. Reputable computer programs should take account of the buoyancy lost when a particular compartment is flooded, and should enable calculation of the flotation and stability with the compartment flooded level with the sea, that is in the equilibrium flooded state, or with some fixed volume of floodwater in the compartment.

The effect of flooding on stability may be complex and difficult to predict. Typically the stability will be reduced, but it may be increased because the associated sinkage may result in an increase in the residual waterplane inertia, and a significant rise in the VCB. The trim is likely to be affected, and may govern the chances of survival because of its effect on the residual freeboard of openings which may lead to further flooding.

A yacht which is stable when inverted will benefit from an ingress of water in one respect, because its stability will be reduced and it will be more likely to self right.

Interpreting damage, or flooded stability, data

Care must be exercised when interpreting computed damage stability data, to ensure that the quoted displacement and associated GZ value correspond to those of interest. When damaged, the floodwater is able to pass through the hull freely, and the quantity therefore may depend on the heel angle. It is normal to assess damage stability on the basis of the displacement of the vessel prior to flooding.

It is likely that the computer program will calculate the righting moment and obtain the GZ value through division by the displacement. If the calculation method involves the use of a displacement which has been adjusted for some added floodwater, the presented value of GZ may correspond to this increased displacement and hence be lower than the value which would be associated with the original unflooded displacement. Sometimes, where flooding affects a compartment containing a tank, the intact volume of the tank may be taken account of simply by subtracting its contents from the loading condition and assuming that it too is flooded. The displacement used in such a calculation will be less than the original displacement, and the presented GZ value, if not readjusted, will be correspondingly high. The method used and the values presented will depend upon the software and how it has been used, and the nature of the result may not be clear from the presentation.

CHAPTER 4
AERODYNAMICS OF SAILS

J. Wellicome
University of Southampton

4.1 INTRODUCTION

The performance of a sailing yacht results from a balance of hydrodynamic forces acting on the hull and aerodynamic forces acting on the sails. This balance of forces is described in Chapter 2. The present chapter is devoted to a discussion of the generation of forces on the sails. Sails are examples of lifting foils and the mechanisms of force generation on each are similar for the whole family of lifting foils, including such items as aircraft wings, gas turbine blades, ship's propellers, cooling fans and many other devices. In many cases the design requirement for a lifting foil is to generate a given lift force while incurring the smallest possible drag penalty. This is certainly the requirement for a yacht keel; however the requirement for the sails is (more nearly) to generate a maximum driving force without incurring excessive side force or heeling moment. This results in a requirement to operate the sails at higher levels of lift, closer to the point of maximum lift than would be the case of most lifting foils. Further features of sails that cause them to differ from most foils are their flexible geometry and the fact that there are usually two or more sails in close proximity. Thus sails are different from most other lifting foils and require a different kind of design methodology.

This chapter presents a discussion of the nature of the airflow round sails and the mechanisms of force generation, the nature of sail interactions, the relationship between sail geometry and sail forces and the need to be able to adjust geometry. Typical wind-tunnel model test data are also presented.

4.2 AIRFLOWS AROUND THICK AND THIN AEROFOILS

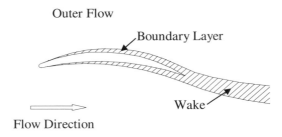

Figure 4.1 *Outer flow, boundary layer and wake regions round an aerofoil: normal attached flow.*

The fluid flow past objects such as lifting foils, operating in either air or water, can be broken into two regions: an outer region in which fluid frictional effects (associated with fluid viscosity) are negligibly small and a region extending over the surface of the foil and downstream of the trailing edge in which frictional effects are important. Over the foil this region is called the boundary layer while downstream of the foil the region is called the wake. The boundary layer is usually quite thin: at the trailing edge the thickness of the layer is typically of the order 3–5% of the chord length of the foil. The forces acting on the foil depend on the way in which this boundary layer develops along the foil chord. This in turn depends on the pressure distribution and, in particular the pressure gradients, across the chord. The pressure distribution is largely determined by the flow outside the boundary layer. A region of falling pressure is said to be a region of favourable pressure gradient, while a region of rising pressure represents an adverse gradient. The reasons for this designation will appear shortly.

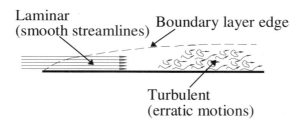

Figure 4.2 *Laminar and turbulent flows.*

Near the leading edge of a foil the flow within the boundary layer may proceed in a smooth, orderly, streamlined fashion called laminar flow. Further from the leading edge this breaks down into a highly erratic flow in which a mass of small eddies are present in a flow field, with velocities varying very rapidly in space and time. This erratic flow is called turbulent flow. On an average the flow will still be following the foil surface. The change from a laminar flow region to a turbulent one is called transition. A favourable pressure gradient will delay transition to a point further from the foil leading edge, possibly back as far as the point of minimum pressure. In contrast, an adverse gradient will cause almost instantaneous transition.

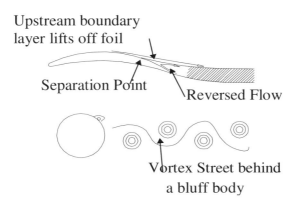

Figure 4.3 *A variety of separation patterns.*

A completely different phenomenon is that of separation. If the pressure gradient is sufficiently adverse the boundary layer will be unable to maintain contact with the foil surface and will lift off, leaving a region of reversed flow between itself and the foil surface. Flow speeds in this separation zone, relative to the foil, are very small (at least in the chord-wise direction). The boundary layer becomes increasingly sensitive to adverse pressure gradients as distance from the leading edge increases. At the trailing edge even a mildly adverse gradient will cause separation.

Transition and separation are quite different concepts. It is possible for a laminar boundary layer flow to separate in a laminar fashion, although an adverse pressure gradient is more likely to trigger transition to turbulence instead. Subsequently separation may occur in the now turbulent boundary layer. Even if the flow is laminar at separation, the flow after the separation point rapidly becomes turbulent.

Figure 4.4 *Leading edge separation bubbles and underside separation on highly cambered foils.*

Near the leading edge a much stronger adverse gradient is required to cause separation, but, remarkably, a return to a favourable gradient can bring about a re-attachment of the boundary layer, thus forming a leading edge separation bubble. This is especially a feature of flows round thin aerofoils. It does not adversely affect foil performance.

Normally, pressure gradients are more intense and more likely to be adverse on the upper surface (or suction surface) of the aerofoil. Separation is most frequently found on the suction surface. However, a feature of thin, highly cambered foils is that a separation bubble can be found on the lower (or pressure) face of the foil. Again, within such a separation bubble, the fluid velocity component parallel to the chord is very small at the foil surface. The upper surface of a foil corresponds to the leeward side of a sail and the lower surface to the windward side. Separation bubbles will exist on each side of a sail set behind a mast.

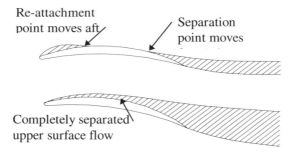

Figure 4.5 *The merging of leading edge bubble and trailing edge separation in the fully stalled state.*

As lift increases, due either to increasing angle of attack to the flow or by increasing foil camber, the pressure gradients on the suction surface of the foil become more severe. The separation point for the trailing edge separation zone moves rapidly towards the leading edge, while a leading edge bubble lengthens. A point is reached when the two zones merge and the entire upper surface flow lifts off the foil surface. During this process the lift rises to a maximum and then reduces. Simultaneously the drag rises sharply. This process is referred to as stalling the foil. Generally, sails designed for upwind performance operate below stall when sailing close hauled, near the point of maximum lift when close reaching

Aerodynamics of Sails 31

and in a stalled state off wind. The reasons for this state of affairs will be made clear later.

4.3 MODERN AEROFOIL DESIGN METHODS

Modern aerofoil design methods involve choosing a pressure distribution that will lead to a proper boundary layer growth along the foil. This involves optimising the extent of laminar flow over the suction surface by choosing a pressure distribution that has a gentle favourable pressure gradient to a point well aft of the leading edge and by limiting the adverse gradient over the rear part of the foil so as to delay the separation point as far as possible. Theoretical treatment of the boundary layer flow leads to criteria for the occurrence of both transition and separation. Having determined a target pressure distribution the design process involves a mathematical derivation of the foil shape which would achieve the target distribution in its design state. NACA sections, which are very popular in yachting circles, were designed by this route using purely inviscid methods of relating pressure to foil shape. Up-to-date design tools (such as the XFOIL code) match inviscid outer flow calculations to calculations of viscous boundary layer growth. Up to the point of maximum lift, lift and drag predicted by these codes compare closely with experimental values. These design tools work very well, provided the foil is not operating too close to its maximum lift; they completely fail once the stalling process gets underway (as the author discovered when trying to design sections for use on stall-regulated wind turbines!). Details of these methods are not of direct concern here.

4.4 THE RELATION OF PRESSURE DISTRIBUTION TO FOIL GEOMETRY

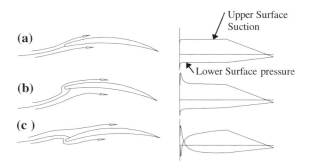

Figure 4.6 *Thin rigid foil streamlines & pressure distributions (**a**) at ideal incidence, (**b**) above ideal incidence and (**c**) below ideal incidence.*

The pressure distribution round a foil is a function of three things: the shape of the camber line, the distribution of thickness across the chord and the angle of attack between the foil nose-tail datum line and the relative airflow direction. According to thin aerofoil theory the pressure distributions due to these effects are linearly additive. For any given camber line shape it is possible to identify a specific angle of attack, called the ideal angle of attack, for which the fluid flow divides smoothly either side of the leading edge of the foil. As the angle of attack increases above the ideal angle the point of attachment of the dividing streamline moves onto the pressure face of the foil, resulting in high flow speeds round the leading edge and a sharp suction peak on the suction side of the foil in this region. The pressure rises again almost immediately, causing a severely adverse pressure gradient just behind the leading edge. The thinner the foil the more severe the adverse gradient. It is this gradient that is responsible, initially, for the formation of the leading edge separation bubble, causing early transition to turbulent flow and, subsequently, contributing to the stalling process. If the angle of attack reduces below the ideal angle the dividing streamline moves onto the suction surface and the pressure peak is on the pressure surface, with corresponding separation effects on the flow over that surface of the foil.

Foil drag is usually at a minimum when operating close to the ideal angle of attack, especially for foils designed to have extensive laminar flow. A sharp rise in drag can occur when leading edge pressure gradients cause transition to move to the leading edge of either the suction or the pressure surface. The ideal angle of attack increases as the camber of the foil is increased, and so, correspondingly, does the lift generated at the ideal angle. The range of angles of attack, on either side of the ideal angle, for which foil performance is satisfactory, is dependent on the radius of the foil leading edge, and thicker foils are more tolerant of variations in angle of attack. Sails are the ultimate thin aerofoil and, as such, should be operated close to their ideal angle of attack. This requires the adjustment of sail camber (or fullness) as the sail lift requirement changes. Hence bendy masts and other sail flattening devices. As a further point, the flexibility of normal sail materials allows the camber line of the sail to change shape as angle of attack changes. Most obviously, as a suction peak develops on the windward (pressure) side at low angles of attack the luff lifts to windward. These changes of shape blur the meaning of ideal incidence for a sail.

The lift and drag characteristics of a typical aerofoil section can be related to the changes in flow regime discussed so far. It can clearly be useful to identify the various flow states on a working sail as a way of ensuring proper sail settings. Wool tufts near the luff and along the leach are the most common ways of doing this.

Lift and drag curves for a NACA section 64–208 (which is, relatively speaking, thin and highly cambered) are shown in Figure 4.7. For this section the camber line has an ideal angle of attack of zero degrees and an ideal lift coefficient of $18.1 \, m/c$, where m/c is the camber/chord ratio.

Aerodynamics of Sails 33

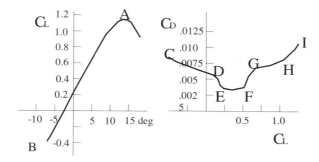

Figure 4.7 *The lift and drag characteristics of a foil.*

A   Point of maximum lift coefficient – suction surface separation point moving rapidly forward.

B   Point of minimum lift coefficient – pressure surface separation point moving rapidly forward.

C–D   Lower surface boundary layer turbulent from leading edge.

E–F   Laminar drag bucket – extensive laminar flow over both foil surfaces (only for smooth foils)

G–H   Upper surface boundary layer turbulent from leading edge.

H–I   Rapid drag rise as upper surface flow separates.

4.5 THREE DIMENSIONAL EFFECTS ON LIFTING FOILS

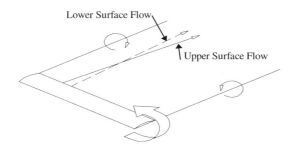

Figure 4.8 *Tip vortex formation.*

So far the discussion has implicitly related to two-dimensional foil properties which would describe flow behaviour at mid span of a high aspect ratio foil. That is, a foil whose span is far larger than its chord. For a foil of finite span there are three-dimensional effects associated with pressure equalisation round the tips of the foil. On the pressure face of the foil the result of pressure falling off towards the tips causes the flow across this face to be deflected outwards towards the tips. On the suction face the reverse is true: pressure is lowest at mid-span and the flow is drawn inwards towards mid span. The effect is strongest near the tips themselves. The flow round the tips from pressure face to suction face results in the formation of a pair of tip vortices, as shown in Figure 4.8, which extend into the wake downstream of the foil. The flow generated by these vortices produces a downwash over the foil and the wake downstream. This is a flow perpendicular to the foil and wake directed from the suction side to the pressure side. Outside the foil/wake area the flow induced by these vortices is an upwash from pressure side to suction side, although this effect diminishes rapidly as distance from the foil/wake sheet increases. The formation of trailing vortices is not confined to the tips, but spreads across the span of the foil.

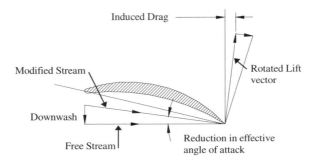

Figure 4.9 *Reduction in lift and the creation of induced drag due to downwash.*

The downwash produced by the trailing vortices has the effect of rotating the fluid flow onto the foil and of rotating the lift and drag axes too. This rotation reduces the effective angle of attack of the foil and hence reduces the lift to a significantly lower level than a two-dimensional section produces. Moreover, the rotation of the lift direction introduces a component of force in the downstream direction that is equivalent to an additional component of drag. This drag component is called induced drag. The viscous drag of a two-dimensional foil is known as profile drag and the drag of a finite span foil is the sum of the profile and induced drags. For a finite span foil operating at a maximum lift/drag ratio the profile and induced drags are equal. For foils, such as sails, operating at high lift the induced drag is much the larger drag component.

Many factors contribute to the distribution of lift downwash and induced drag across the span of a foil:

- foil aspect ratio (span/mean chord)
- foil planform (spanwise variation of chord)
- spanwise twist distribution.

Note that for a sail system in a near vertical attitude the equivalent of a wing downwash is a flow to windward and upwash a flow to leeward.

4.6 IDENTIFYING INDUCED DRAG

Since downwash increases with the strength of the trailing vortices, which in turn increase as lift increases, the angle of rotation of the lift increases with lift and hence the component of lift in the downstream direction increases as the square of the lift. It is common practice to plot a diagram of foil drag to a base of the square of lift. At least until stall commences this graph is more or less a straight line whose intercept on the zero lift axis indicates the profile drag of the foil and whose slope indicates the induced drag.

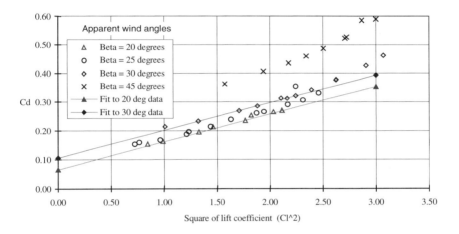

Figure 4.10 *Lift and drag coefficients for main with masthead genoa, including windage.*

A diagram such as Figure 4.10 may be more meaningful for a sail configuration if the tare lift and drag of the hull topsides, masts and rigging have been removed from the total aerodynamic loads first. In fact this diagram may be drawn with or without the tare contribution. This should be borne in mind when looking at such a diagram. The right-hand end of Figure 4.10 shows clearly the rapid drag rise and the loss of lift as stalling occurs.

A typical variation of the tare contribution is shown in Figure 4.11 in which the reference sail area has been used to compute Cl and Cd.

36 Sailing Yacht Design: Theory

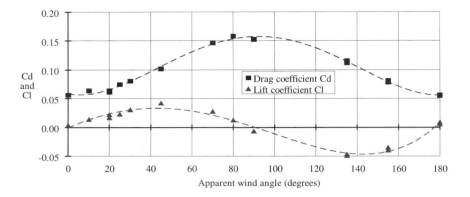

Figure 4.11 *Lift and drag coefficients for windage of hull, superstructure and rig, based on sail area.*

4.7 IDENTIFYING THE MAXIMUM FORWARD DRIVE FORCE

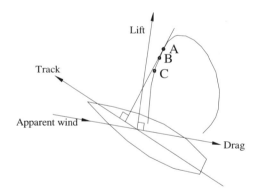

Figure 4.12 *Sail polar set relative to apparent wind axis.*

Another useful plot is to show the total aerodynamic drag (for the whole yacht) plotted against the total lift. Such a curve should represent an envelope of performance enclosing results for a wide range of sheeting angles, the curve being drawn round the data on the high lift/low drag side. If this plot is lined up so that the drag axis is along the apparent wind direction as shown in Figure 4.12, the aerodynamic force vector for any point on this envelope can be resolved into a drive force component parallel to the track of the yacht, and a side-force component. The side-force component will be associated with a heeling moment which must not be excessive. It will be resisted by the hull side-force, which in turn will increase the drag of the hull. A point on the envelope can be identified corresponding to the maximum forward drive force (A). Were it not for hull side-force induced drag effects, operating the sails at this maximum forward drive condition would produce maximum boat speed. In practice the best boat speed is obtained using slightly less lift, as at point B (say), and this represents the best light weather sail setting. In heavy weather it will be necessary to reduce sail forces further, to (say) point C, in order to limit heeling. Of course, the points A, B, C are not unique: they all move as the apparent wind angle changes.

In light weather, while reaching, sails should be operated near to the point of maximum drive force, itself close to the point of maximum lift. At this condition some trailing edge separation will be present i.e. the sails are moving towards a stalled state. Furthermore, in the case of a foresail there will probably be a leading edge separation bubble present on the lee side of the sail.

When close-hauled a lower level of lift should be used, and in this state the flow will be fully attached, except locally behind the masts. Typical lift coefficients when reaching are about 1.4–1.6 while close-hauled lift coefficients are about 1.0–1.2.

4.8 OPTIMUM PLANFORM AND TWIST FOR A SINGLE SAIL

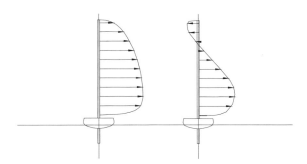

Figure 4.13 *Optimum spanwise sail loadings in light and heavy weather.*

Most lifting foils are designed to generate a certain lift while incurring the smallest possible drag. This requires that induced drag should be reduced as far as possible. High downwash implies a high induced drag for a given lift. By redesigning a foil planform or its spanwise distribution of twist, it is possible to transfer lift across the span from a region of high downwash to a region of low downwash with a consequent reduction of induced drag, provided the downwash is not uniform across the span. Thus, in terms of minimising induced drag, a foil with a non-uniform downwash distribution is non-optimum and, conversely, an optimum foil requires uniform downwash. For an untwisted foil operating in a uniform flow a foil with an elliptic planform produces a uniform downwash. It is for this reason that the elliptic planform, typified by the Spitfire wing, is usually regarded as the ideal. However, the maximum lift/drag ratio criterion is not the proper one for a sail. The sail requirement is to produce maximum boat speed on a given course in the prevailing wind strength. In light weather this almost corresponds to producing the maximum possible driving force (provided the side-force involved does not result in excessive keel induced drag). This is not quite the same as producing uniform downwash, but is not very much different. In heavy weather, at least working to windward, the requirement is to produce the maximum boat speed subject to the requirement to limit heel angle. This is more or less the same as producing the maximum driving force within a fixed maximum heeling moment. A sail which generates a lift to windward at the masthead can generate much larger driving force lower down the sail and theoretical studies have shown the spanwise loading which maximises forward drive force within a given heeling moment needs to produce a substantial windward lift high up the sail. The required planform is a long way from elliptical and the sail requires considerable twist near the masthead to achieve the required lift distribution. In fact the traditional triangular Bermudan sail is not as wrong as sailing folk have been led to believe for the last forty years or so!

In terms of sail twist there are several complications to bear in mind:

- The sail is not operating in a uniform airflow. The natural wind varies in speed and direction with height from the sea surface. In combination with the forward motion of the yacht at a given course angle to the wind, the resulting apparent wind will vary in direction up the mast. It will be different on port and starboard tacks due the twist of the natural wind component.
- The airflow is modified by the hull topsides. This removes some of the twist in the apparent wind direction working to windward and augments it off the wind.
- The wake from each sail produces a pattern of downwash for all sails astern (and upwash for all sails ahead) which varies as the spacing between the sails changes when the sheets are adjusted to suit different courses to the wind.

All this sounds complicated enough. However, the practical outcome is that there is a need to be able to control the twist of a sail as well as its camber in order to get the best out of the rig. As a general rule a mainsail should be nearly twist free when working to windward in light weather, while quite substantial twist is appropriate off wind.

4.9 THE EFFECT OF A GAP AT THE SAIL FOOT

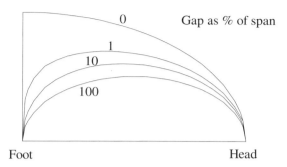

Figure 4.14 *Optimum loading with gap.*

The effect of a gap between the foot of a sail and the deck can be conveniently modelled by a rig mounted above a flat ground board. The ground board itself can be replaced by a second rig that is the reflection of the true rig about the plane of the ground board. In the wind tunnel the rig plus its reflection produces a flow similar to that of the rig above a ground plane. The upwash over the rig produced by the image of the rig modifies the downwash distribution over the rig itself. In turn this modifies the spanwise loading that produces the optimum downwash distribution. For a large gap the effect of the image is weak and the optimum loading in light weather is nearly elliptical from foot to head. As the gap reduces to something small the optimum loading approaches elliptical over the whole rig plus image system. The sail load is still required to be zero at the sail foot as shown in Figure 4.14.

4.10 THREE-DIMENSIONAL EFFECTS IN SEPARATED FLOW REGIONS

Inside a separation zone flow speed across the chord of the foil is small and possibly reversed in direction. If the planform is swept backwards or forwards a noticeable spanwise flow is observed in a separation zone directed towards the trailing part of the foil. Thus, in a bubble separation zone along the leading edge of a foresail there is a flow upwards towards the masthead. In the trailing edge separation zone the flow is downwards towards the sail foot. If there is a separation zone in the middle of the sail on the weather side the flow in that zone will also be downwards. The direction in which a wool tuft streams can, therefore, be a very sensitive guide to the local flow regime. In a trailing edge separation zone the flow will be unsteady and the tuft will flutter.

40 *Sailing Yacht Design: Theory*

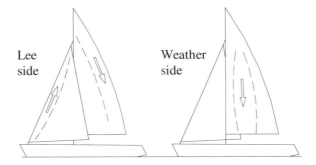

Figure 4.15 *Vertical flows inside separation zones.*

4.11 SAIL INTERACTIONS

Where two or more sails are set in close proximity they will interact in two ways: (a) there will be a pattern of downwash/upwash generated by each sail and the various individual sails will operate in an airflow modified by the presence of the other sails, and (b) if the gap between the sails is sufficiently small the leading sail guides the flow on to the leading edge of the trailing sail and can modify the pattern of separation on the trailing sail.

Of the two types of interaction, the first is always present and will be the dominant effect. The second type will only be significant if separation is already present on the trailing sail. Curiously, the proper way to use this slot effect [type (b)] is not to eliminate separation, but to increase the sail lift that can be generated before separation becomes excessive. This is because, at least in light weather, the proper sail setting is near the point of maximum lift at which the stall process has already started. Of course, when close-hauled, the sail is operating well below stall and the slot effect is minimal.

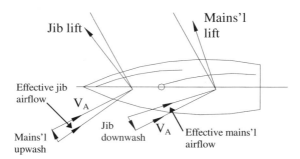

Figure 4.16 *The effect of downwash/upwash on forward drive force.*

Ahead of a sail there is a region of upwash. The upwash from a trailing sail increases the apparent wind angle of a leading sail. The consequent rotation of the lift vector of the leading sail increases its forward drive force for a given level of side force, thus improving its performance. A leading sail has the reverse effect on a trailing one. Thus a foresail is more effective than the mainsail behind it, while a mizzen is relatively ineffective because of the presence of the fore and main sails ahead. This is illustrated in Figure 4.16.

The downwash/upwash will modify the sheeting angles required of each sail. Prime examples of this being the need to sheet the main boom to windward in a 1970s IOR masthead rig with the largest Genoa set and the progressive changes of yard angle observed in overhead photographs of multi-masted square riggers.

4.12 THE EFFECT OF HEEL ON SAIL PERFORMANCE

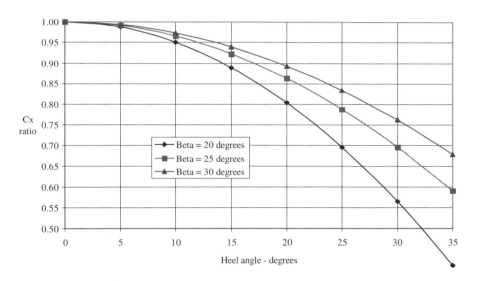

Figure 4.17 *Variation of driving force ratio, upright to heeled, with heel angle.*

As the yacht heels over the sail forces reduce, the effect being more pronounced on the sail driving force than on the heeling moment. The major part of this effect can be estimated by resolving the apparent wind vector into a component parallel to the mast and a component perpendicular to the mast. It is this latter component that the sail sees as the effective onset flow, which determines the heeled sail forces. There is a reduction of the effective apparent wind angle and speed that reduces the sail angle of attack and sail forces, in particular the driving force component, as shown in Figure 4.17.

4.13 REACHING AND DOWNWIND SAILS

Once the apparent wind angle opens up beyond, say, 40°, it is usual to set special off-wind sails from large reaching genoas and conventional spinnakers to a variety of hybrid sails that are a cross between these two. There is a certain amount of wind tunnel data for these sail types that enables comparisons to be made between them. The following diagrams are taken from reference 1.

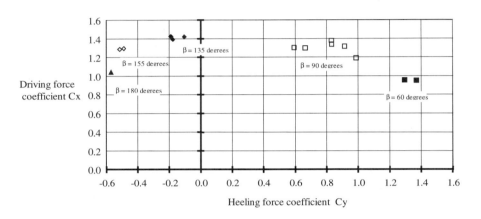

Figure 4.18 *Variation of driving force coefficients with heeling force coefficients, from USS/PHRF tests with fractional spinnaker, excluding windage.*

Figure 4.18 shows driving force and heeling force coefficients for a fractional spinnaker. As with all sail coefficients, the forces are reduced to coefficient form using a nominal sail area that may be defined in a variety of ways. The nominal area for Figure 4.18 was taken as 87% of the product of the luff and mid girth lengths for the sail. On this basis the spinnaker forces are very similar to those for a simple mainsail plus genoa combination. The forces shown are components of the total aerodynamic force in a horizontal plane. In the case of balloon-like sails there can be an additional vertical force component.

Figure 4.19 is a similar diagram for a variety of reaching headsails set in combination with a mainsail. Again, provided the appropriate nominal sail area is used for each sail combination, the performance of all the sails is comparable. Bearing in mind the earlier remarks about ideal angles of attack, getting the best performance out of a sail at a given apparent wind angle requires the proper choice of sail camber. This needs to be increased as apparent wind angles increase off-wind. Given the correct sail geometry, there can be remarkably little flow separation over reaching sails and spinnakers, even near running conditions.

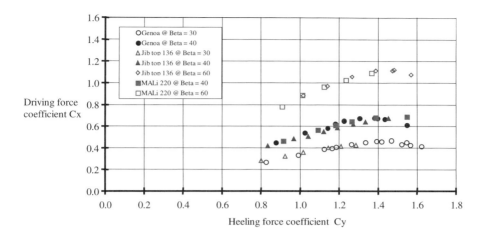

Figure 4.19 *Variation of driving force coefficient with heeling force coefficient, from W60 tests including hull windage.*

4.14 CENTRES OF EFFORT OF A SAIL PLAN

A convenient way of defining the centre of effort of a sail plan is to divide the heeling moment and the yawing moment acting on the vessel by the heeling force component. This defines the point at which the aerodynamic force cuts the centreline plane of the yacht. Both the height and longitudinal position of the centre of effort vary with sail adjustments such as sheeting angles and kicking straps etc. to control sail angles of attack, twist and camber.

Figure 4.20, shows centre of effort heights approximately equal to the height of the geometric centre of area of the sail plan. However, changing sail twist can result in a significant vertical movement of the aerodynamic centre.

44 Sailing Yacht Design: Theory

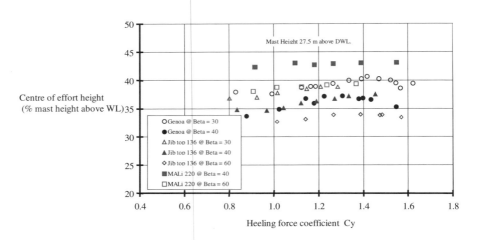

Figure 4.20 *Variation of centre of effort height with heeling force coefficient, including hull windage.*[2]

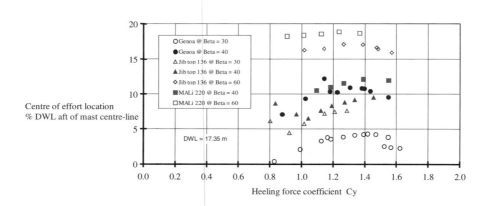

Figure 4.21 *Variation of centre of effort location with heeling force coefficient, including hull windage.*[2]

Figure 4.21, shows the longitudinal position of the centre of effort of the rig. This is much more variable than the vertical height and sensitive to individual sail trimming. As an incidental comment, the distribution of sail camber from luff to leach has a major influence, moving aft as the deepest part of the sail is moved towards the leach. This can be used to balance the rig of single sailed una-rig dinghies by flattening the luff of the sail using the Cunningham eye.

REFERENCES

1. Campbell, I.M.C. *The performance of offwind sails obtained from wind tunnel tests*. RINA, Int. Conf. on The Modern Yacht, Portsmouth, March 1998.
2. Campbell, I.M.C. *Optimisation of a sailing rig using wind tunnel data*. SNAME 13th Chesapeake Sailing Yacht Symposium, Anapolis, USA, 1997.
3. Day, A.H. *Sail optimisation for high speed craft*. RINA 1991.
4. Day, A.H. *Steps towards an optimal yacht sailplan*. RINA, 1994.

BIBLIOGRAPHY

Marchaj, C.A., *Sailing Theory and Practice*, Adlard Coles Limited, London, 1977.

Marchaj, C.A., *Aero-Hydrodynamics of Sailing*, Adlard Coles Limited, London, 1979.

Milgram, J.H., 'The Aerodynamics of Sails', *Proceedings of the Seventh Symposium on Naval Hydrodynamics*, Office of Naval Research, 1968.

Milgram, J.H., 'Analytical Design of Yacht Sails', *Transactions,* The Society of Naval Architects and Marine Engineers, 1968.

Gentry, A.E., 'The Aerodynamics of Sail Interaction', *Proceedings of the Third AIAA Symposium on the Aero/Hydronautics of Sailing (The Ancient Interface III)*, American Institute of Aeronautics and Astronautics, 1971.

Greeley, David S., *et al.*, 'Scientific Sail Shape Design', *Proceedings of the Ninth Chesapeake Sailing Yacht Symposium*, Society of Naval Architects and Marine Engineers, 1989.

CHAPTER 5
THE HYDRODYNAMICS OF HULL, KEEL AND RUDDER

J. A. Keuning
Delft University of Technology

5.1 INTRODUCTION

The total hydrodynamic resistance of a sailing yacht hull underway may be divided into several components. Most of them are identical to those of a 'normal' vessel. These are: the frictional resistance, the form drag (both due to the viscosity of the fluid) and finally the wave resistance, which originates from the presence of the free water surface and the waves generated therein by the moving hull. In addition to these however the sailing yacht hull experiences some extra resistance forces in its regular steady sailing condition, which may be attributed to the fact that the hull generates sideforce to withstand the lateral components of the sail forces : i.e. the induced resistance and the resistance due to heeling of the hull.

Due to the viscosity the particles immediately adjacent to the body surface come to a complete stop and the fluid particle velocity around the body, as predicted by the potential theory, is only reached a certain distance away from the hull. This results in a layer of water which gets gradually thicker when moving along the length of the body from bow to stern, leaving an area of retarded flow, in which the fluid particle speed varies from zero to the potential flow pattern velocity and this is known as the 'boundary layer'.

Shear forces in the boundary layer and in particular on the body surface are caused by the viscosity and the velocity gradient in the boundary layer giving rise to frictional resistance.

In addition to the frictional effect there is a pressure deficit over the afterbody and this yields a force on the body in the direction of the undisturbed fluid velocity which we all know as 'resistance'.

So a fully submerged body in a stationary condition experiences two types of resistance both of which are viscous in nature, i.e. the frictional resistance and the eddy making resistance or form drag.

When the body moves towards the free water surface an additional resistance component develops: the wave-making resistance. A ship moving in the free surface experiences the frictional resistance and the form drag just as a fully submerged body does. However the movement of the ship through the water causes a pressure distribution around the ship similar to the one found around the fully submerged body, i.e. there is an area of increased pressure near the bow and the stern of the ship and an area of reduced pressure at the middle. This variation in pressure in and just below the free surface causes a wave disturbance in the free surface, because this free surface is a plane of equal (atmospheric) pressure. Changes in pressure have to be compensated by a rise or fall of the free water surface. So waves are continuously being developed which travel away from and behind the ship. The energy in this wave system has to be drained from the ship and wave-making resistance arises as an additional resistance component.

These surface waves originate from the pressure distribution along the length of the moving hull in the free surface and may be calculated by using potential theory for ideal fluids. However the actual system around the real ship is influenced by the viscosity of the fluid. The existence of the boundary layer influences the pressure distribution along the hull and therefore the waves being generated, and also the friction and form drag is influenced by the surface waves. The final result is a very complicated interaction between all the various components.

So far the resistance components of a 'normal' ship and a sailing yacht coincide. A sailing yacht however experiences additional forces on the underwater part of the hull due to the fact that the sailforces acting on the above water part of the hull have to be counteracted. These forces may be divided into lateral forces and resistance forces.

As explained in Chapter 2 the hull will heel over due to the combined action of the aerodynamic and hydrodynamic forces. Due to this angle of heel the underwater part of the hull changes from symmetric to asymmetric and also a change in wetted area of the hull may occur. The latter will result in a change in the frictional resistance. In addition, however, the form drag of the hull can change due to the asymmetry of the flow pattern around the heeled hull.

The waterline length, the waterline beam, the canoe body draft and the distribution of the volume of the hull over the length of the hull may also change dependent of the particular geometry of the yacht, as shown in Figure 5.1, and this will lead to a change in the wave-making resistance of the hull.

The last resistance component of a sailing yacht hull is directly related to the sideforce generated on the hull and appendages and is called induced resistance. Since the keel and rudder of a sailing yacht and to a lesser extent the hull itself, generate lift, or sideforce, to counteract the sideforce generated by the sails and, since the span of these wings may be expected to be rather limited, a sailing yacht hull will experience in most cases considerable induced resistance. This induced resistance is dependent on the specific geometry and layout of the appendages and the prevailing sailing conditions of the yacht under consideration.

48 *Sailing Yacht Design: Theory*

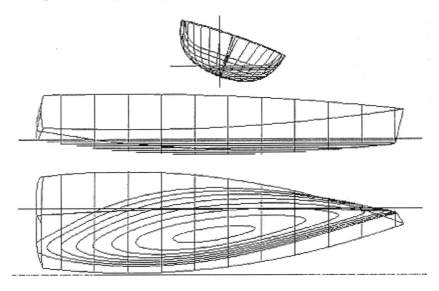

Figure 5.1 *Lines plan of a heeled canoe body.*

Each of the above mentioned resistance components plays an important role in the performance of a sailing yacht. In the following sections each of them will be described in more detail and possible approaches to estimate their magnitude will be discussed.

5.2 VISCOUS RESISTANCE

The frictional resistance

The viscous resistance of a yacht hull is, according to the explanation presented above, divided in two separate parts: the frictional resistance and the form drag or eddy making resistance.

The frictional resistance is found to be dependent on the area of the surface of the hull in contact with the water, i.e. the wetted surface of the hull (S_c), the forward speed of the yacht (V_s^2) and a frictional coefficient (C_f). It may then be formulated according to expression [5.1]

$$R_{friction} = \tfrac{1}{2}\,\rho V^2\,C_f\,S_c \qquad [5.1]$$

The determination of this friction coefficient C_f is based on the results of experiments in towing tanks and wind tunnels with flat plates in the direction of the flow, which are therefore supposed to have no form drag and wave resistance and so all the measured resistance is frictional resistance. These tests were first carried out by Froude in 1872 and repeated thereafter by many researchers in various research establishments all over the world. Froude carried out his experiments with flat plates of various lengths with a smooth surface. He found the resistance per unit of area to be dependent on the length of the plates in such a way that the longer plates had a lower resistance per unit of area.

After the work of Reynolds this difference in the specific resistance was found to be related to changes of the quantity VL/ν, known hereafter as the Reynolds number (R_N). (V = velocity, L = length, ν = kinematic velocity.)

A typical plot of the frictional coefficient Cf, as found by Blasius for laminar flow and by Prandtl for fully turbulent flow is presented in the Figure 5.2 as a function of this Reynolds number. The difference between the two is obvious.

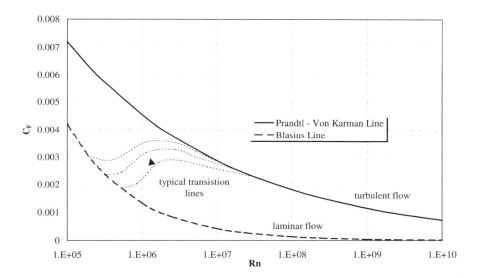

Figure 5.2 *Blasius laminar flow line to Prandtl turbulent flow line transition.*

It should be noted however that from the experiments carried out by Reynolds it became obvious that a laminar flow may no longer be feasible at or above a certain 'critical' magnitude of the Reynolds number, Rc. Above this Rc the flow becomes turbulent. He found this transition from laminar to turbulent flow to take place around $R_N = 4.5 \times 10^5$. This is not a hard limit because the actual transition appeared to be strongly dependent on the degree of background turbulence of the water, the roughness of the plate's surface and some of the plate's dimensions.

To unify the magnitude of the Cf used in towing tanks around the world, the International Towing Tank Conference (ITTC) decided in Madrid in 1957 to adopt one single formulation for the purpose of extrapolation of model to full scale resistance, which became known as the ITTC–57 friction line:

50 *Sailing Yacht Design: Theory*

$$Cf = \frac{0.075}{\left(Log(R_N) - 2\right)^2} \qquad [5.2]$$

For 'normal' ships the still water waterline length L is used for the calculation of the Reynolds number. For yacht hulls this does not really represent the path of travel of the particles in the actual flow. So some kind of average waterline length is used which tries to take into account these differences. For a typical hull shape as depicted in Figure 5.3a 70% of the waterline length appeared to be a reasonable approximation, for hull shapes like the one in Figure 5.3b 90% of this length is used. The discrimination for an arbitrary hull form leaves some space for interpretation.

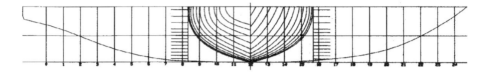

Figure 5.3a *Body plan DSYHS Series 1 parent model.*

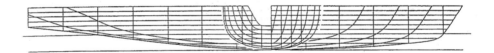

Figure 5.3b *Body plan DSYHS Series 2 and 3 parent model.*

Figure 5.3c *Body plan DSYHS Series 4 parent model.*

The form factor

The actual frictional resistance of a yacht hull will differ somewhat from this approximation used for the calculation because the shape of the wetted surface of the hull is not identical to that of a flat plate. Using the flat plate approximation for the wetted surface of the hull implies a two-dimensional approach to the flow while in reality the flow around the hull is a three-dimensional flow. The difference between the two flow conditions is accounted for by what is generally known as the 'form factor' k. The form factor should depend on the shape of the hull only and not be dependent on the R_N number. The form factor is usually obtained from analysis of the towing tank data from tests with a model of the hull. There are several procedures which may be used for the determination of k from model test results. One approach which is widely applied by the hydrodynamic community is derived from the method as presented originally by Prohaska and is known as the Prohaska plot. The underlying assumption underneath this approach is that at very low speeds the wave-making resistance has a predictable dependency, typically proportional to Fn^4, on forward speed and should become zero when extrapolated to zero forward speed. By this method k also takes account of the form drag or eddy resistance. A typical example of such a Prohaska extrapolation plot is presented in Figure 5.4.

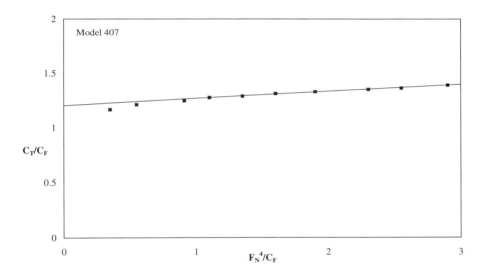

Figure 5.4 *A typical Prohaska extrapolation plot.*

The frictional resistance at each forward speed of the hull is obtained by using the ITTC–57 friction line for the determination of Cf and the wetted surface of the hull at rest. To account for the three-dimensional effects and the form drag this frictional resistance is then multiplied with the factor (1 + k) to obtain the total viscous resistance.

An additional increase in viscous resistance is caused by the effects of hull surface roughness. The friction coefficients presented by the ITTC–57 formulation account for the skin friction of a smooth surface only.

Appendage viscous resistance

For the appendages a slightly different approach is followed. Here the skin friction of the appendage is also approximated using the ITTC–57 formulation of the friction coefficient as a function of Reynolds number. The specific length for the determination of the Reynolds number is now the average chord length if the appendage doesn't have too much taper, i.e. the difference between the chord length at the tip and at the root. If the taper ratio is in the order of 0.6 or lower the span of the appendage may be divided into several 'strips' and for each strip the appropriate average chord length may be used for the determination of the Reynolds number. The skin friction of the strip is now determined by applying the formulation on the strip only. The total skin friction of the appendage is found by summation of the skin friction of all strips.

The form factor of the section profiles of the appendages may be determined using the considerable amount of data available in the literature and obtained largely from wind tunnel tests. For instance in 'Theory of Wing Sections' Abbott[7] presents a large number of lift and resistance characteristics of various sections investigated by the NACA. A very useful approximation of the form drag of a 2-D wing section, such as those most commonly used for yacht appendages, is given by equation [5.3] as presented by Hoerner[13,14] in his books 'Fluid Dynamic Drag' and 'Fluid Dynamic Lift', i.e.:

$$1 + k = 1 + 2(t/c) + 60(t/c)^4 \qquad [5.3]$$

$t =$ aerofoil thickness
$c =$ aerofoil chord length

It should be noted however that except for the ratio of sectional thickness to chord length (t/c) no other geometric characteristics of the section are taken into account. So if a higher level of accuracy is sought for one particular profile it is probably better to use experimental data, or CFD results as discussed in *Sailing Yacht Design: Practice* – Chapter 9.

5.3 WAVE-MAKING RESISTANCE

The wave pattern being generated

In the introduction to this chapter it is stated that the wave-making resistance which a surface ship experiences when it sails in the free surface is the net force on the hull in the longitudinal direction due to the fluid pressures acting normally on the hull surface.

A first approach to develop an understanding of the waves that are formed by a moving ship was presented by Kelvin at the beginning of this century. He considered the ship to be a moving pressure point in the free surface. A pressure point sends out waves in the free surface in all directions. Due to the velocity of these 'free' waves and the forward velocity of the pressure point itself an interference pattern of all these combined wave systems arises because at one point these waves cancel each other out and at another point they will intensify. The wave pattern that so develops is generally known as the 'Kelvin wave pattern' and is depicted in Figure 5.5.

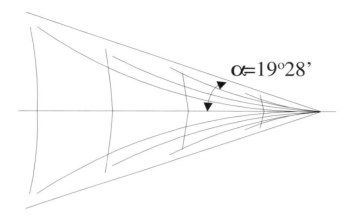

Figure 5.5 *Typical Kelvin surface wave pattern.*

The pattern consists of a series of transverse waves travelling behind the moving pressure point and a series of divergent waves radiating from the pressure point. In deep water the whole wave system is contained in between two straight lines which make an angle of 19° and 28 minutes to the direction of travel of the pressure point. The height of the successive transverse waves diminish with increasing distance from the pressure point. The waves are curved back some distance out from the centreline and meet the divergent waves in 'cusps', which are the highest points in the final wave pattern. The heights of these 'cusps' diminish less rapidly with increasing distance from the pressure point when compared with the height of the transverse wave system and therefore it remains the most prominent or visible aspect of the wave pattern after the ship has moved by.

The wave pattern around a moving ship is somewhat more complicated because it may be considered to be composed by the wave patterns arising from several 'pressure points' of which the bow wave will be the most prominent but also stern wave and the shoulder wave systems are present (but not always easily distinguishable, as they are masked by the bow wave system). The waves are 'free' surface waves for which the relation between wavelength (λ) and speed of advance in deep water is given by equation [5.4]:

$$\lambda = 2\pi V^2 / g \qquad [5.4]$$

Because the wave system is travelling with the forward speed of the ship itself the wave length is dependent on the ship speed: i.e. the higher the speed the longer the waves. The analogy with the well known Froude number is obvious. An example of how the wave length increases with the ship speed and how this positions the wave crest and trough along the length of the ship is presented in Figure 5.6, in which for different Froude numbers the wavelength is related to the ship length. It can be seen that at Fn = 0.40 the wave length equals the ship length and in this situation the crest of the bow wave coincides with the crest of the stern wave. This will result in a considerable increase of the resultant wave height due to the effect of mutual interference at this particular speed which is generally known as the 'hull speed'.

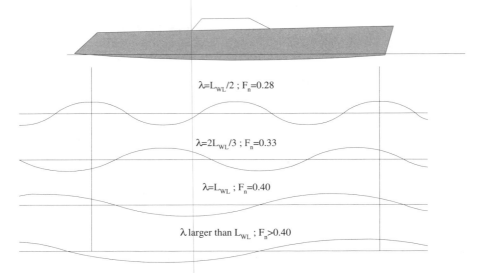

Figure 5.6 *Influence of ship speed on encountered wave length.*

Since the wavelength of the waves is related to the speed of advance of the ship and because different systems originate from different locations along the length of the ship, all these different wave systems will interfere with each other and the overall effect of this on the overall wave height will depend on the ship speed. At certain speeds this will lead to an increase of the resultant wave height and at other speeds they will cancel each other out.

At low speeds the waves which are generated by the ship are generally small, i.e. short in length and low in amplitude. The overall ship resistance will be dominated by the frictional resistance and the resistance will therefore increase with the forward speed squared. As the speed of the ship increases the increase in wave length and the different positions of the crests and troughs will lead to differences in the interference between the different wave systems. This will result in a non monatonic increase of resistance of the ship, so 'humps' and 'hollows' in the resistance curve of the ship arise. A typical result of this is shown in Figure 5.7.

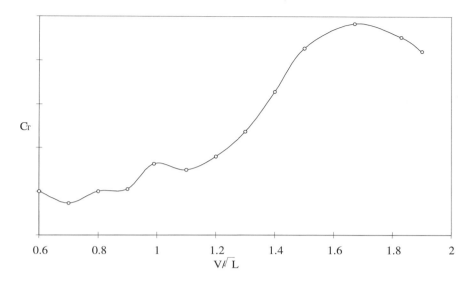

Figure 5.7 *Humps and hollows of a ship resistance curve. (C_T = total resistance coefficient)*

This has become particularly clear from experiments with 'systematic' model series. These are series of towing tank test models all derived from one particular model, the so-called 'parent' model. By changing one parameter at the time while keeping the others as much as possible constant, the influence of that particular parameter on the resistance can be determined by comparison with the parent and the other models. By doing this for a number of combinations of parameters an empirical formulation for the determination of the resistance of an arbitrary hull shape may be obtained.

A number of these systematic series have been tested over the years to give insight in the parameters influencing the wave-making resistance. A notable example is the Delft Systematic Yacht Hull Series (DSYHS) which has been tested over the last 25 years at the Delft Ship Hydromechanics Laboratory of the Delft University of Technology (DUT) in The Netherlands. This series is specially aimed at sailing yacht hulls. In the past 25 years more than 50 different systematically varied models have been tested within this DSYHS and the tests comprise both upright resistance tests as well as tests with the models at heeled and yawed conditions at various speeds.

Residuary resistance calculation by polynomials

To derive empirical expressions for the hydrodynamic forces on a sailing yacht hull the Delft Systematic Sailing Yachts Series was set up in 1974 by Gerritsma *et al.* at the Delft Ship Hydromechanics Laboratory. As a parent hull form at that time the Standfast 43 design of Frans Maas (Breskens, The Netherlands) was chosen as a typical representative of the contemporary designs of that era. In the beginning of the 1980s the designs of the sailing yacht had evolved quite a bit from the lines of this Standfast 43 design and therefore it was decided to incorporate a new parent model into the series which was a design of Van de Stadt and Partners (Wormerveer, The Netherlands), the Van de Stadt 40. Finally in 1995 yet another parent model has been introduced into the series: the IMS-40 designed by Sparkman and Stephens (New York, USA). The bodyplans of all these parent models are presented in the Figures 5.3 a, b and c respectively.

From the results and the knowledge gained from other systematic series in the field of Ship Hydromechanics it was known which hull parameters had a significant influence on the wave-making resistance of ships and to limit the amount of models necessary the following parameters were chosen:

Length–Displacement Ratio ($L/\nabla^{1/3}$)	5.0 to 8.0
Length to Beam Ratio	5.0 to 2.8
Beam to Draft Ratio	2.5 to 19
Longitudinal Centre of Buoyancy (% LWL)	0.0 to 8.0 (aft midship)
Prismatic Coefficient	0.52 to 0.60
Cross Section Coefficient	0.646 to 0.777

All models within the DSYHS have been tested in the upright condition in two configurations; one with and one without the appendages. In the case of the appended hull tests one standard keel and rudder has been used throughout the whole series for the sake of consistency.

Although inevitably more than one formulation of a polynomial expression for estimating residuary resistance is possible, all approximating the original data with small differences in accuracy, the following expression shows a satisfactory combination of the desired level of accuracy combined with sufficient 'robustness', i.e. resistance against possible misuse.

In this approach one polynomial is formulated for each of a large number of separate Froude numbers, i.e. for Froude number of Fn = 0.125 to Fn = 0.600 with an increment of 0.05 and the resistance at intermediate speed values are found by means of interpolation to yield a resistance curve.

The polynomial expression for the residuary resistance of the bare hull (without the keel and the rudder) is given in equation [5.5].

$$\frac{R_r}{\nabla_c 10^3 g} = a_0 + \left(a_1 \frac{LCB_{fpp}}{L_{wl}} + a_2 C_p + a_3 \frac{\nabla_c^{2/3}}{A_{wl}} + a_4 \frac{B_{wl}}{L_{wl}} \right) \frac{\nabla_c^{1/3}}{L_{wl}} + \left(a_5 \frac{\nabla_c^{2/3}}{S_c} + a_6 \frac{LCB_{fpp}}{LCF_{fpp}} + a_7 \left(\frac{LCB_{fpp}}{L_{wl}} \right)^2 + a_8 C_p^2 \right) \frac{\nabla_c^{1/3}}{L_{wl}} \quad [5.5]$$

R_r = Residuary resistance [N]
L_{wl} = Length on waterline [m]
B_{wl} = Beam on waterline [m]
C_p = Prismatic coefficient

V_c = Volume of displacement canoe body [m^3]
LCB_{fpp} = Length centre of buoyancy measured from fore perpendicular [m]
LCF_{fpp} = Length centre of flotation measured from fore perpendicular [m]
A_{wl} = Waterplane area [m^2]
S_c = Area of wetted surface canoe body [m^2]
g = Gravitational constant [9.81 m.s^{-2}]

A full set of coefficients for these polynomials at the various Froude numbers is presented in Table 5.1.

Table 5.1 Coefficients of the resistance polynomial (equation [5.5])

F_n	0.10	0.15	0.20	0.25	0.30	
a_0	−0.00086	0.00078	0.00184	0.00353	0.00511	
a_1	−0.08614	−0.47227	−0.47484	−0.35483	−1.07091	
a_2	0.14825	0.43474	0.39465	0.23978	0.79081	
a_3	−0.03150	−0.01571	−0.02258	−0.03606	−0.04614	
a_4	−0.01166	0.00798	0.01015	0.01942	0.02809	
a_5	0.04291	0.05920	0.08595	0.10624	0.10339	
a_6	−0.01342	−0.00851	−0.00521	−0.00179	0.02247	
a_7	0.09426	0.45002	0.45274	0.31667	0.97514	
a_8	−0.14215	−0.39661	−0.35731	−0.19911	−0.63631	
F_n	0.35	0.40	0.45	0.50	0.55	0.60
a_0	0.00228	−0.00391	−0.01024	−0.02094	0.04623	0.07319
a_1	0.46080	3.33577	2.16435	7.77489	2.38461	2.86817
a_2	−0.53238	−2.71081	−1.18336	−7.06690	−6.67163	−3.16633
a_3	−0.11255	0.03992	0.21775	0.43727	0.63617	0.70241
a_4	0.01238	−0.06918	−0.13107	0.11872	1.06325	1.49509
a_5	−0.02888	−0.39580	−0.34443	−0.14469	2.09008	3.00561
a_6	0.07961	0.24539	0.32340	0.62896	0.96843	0.88750
a_7	−0.53566	−3.52217	−2.42987	−7.90514	−3.08749	2.25063
a_8	0.54354	2.20652	0.63926	5.81590	5.94214	2.88970

To calculate the total resistance of the yacht hull the frictional resistance is added to the residuary resistance using the ITTC–57 calculation method as described in the chapter on the viscous resistance.

From this polynomial approximation of the residuary resistance the dependency of the residuary resistance on some of the hull parameters may be obtained for a constant Froude number.

Because both the prismatic coefficient (Cp) and the Longitudinal position of the Centre of Buoyancy (LCB) appear to the second order in the polynomial expression it may be assumed that both of them have an optimum value as a function of the speed under consideration.

To demonstrate the range of applicability of such a polynomial approximation for assessing the resistance of a variety of yacht hulls the results are shown of a comparison between calculated and measured resistance of a design not belonging to the DSYHS. This design strongly resembles a modern BOC yacht (Figure 5.8). It should be noted that the design is quite remote from the general lines of the three parent models as used in the DSYHS and yet the correlation between the measured and calculated results is quite good.

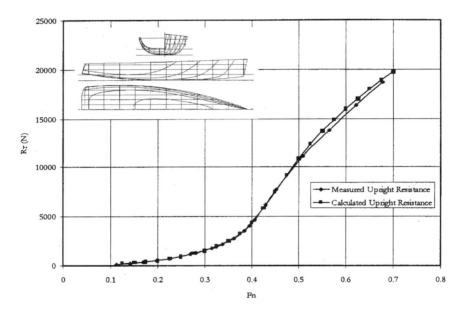

Figure 5.8 *Calculated (using polynomial expression) and measured upright resistance of a BOC type hull.*

Appendage resistance

Until now only the residuary resistance of the bare hull has been considered but the appendages also cause resistance forces. There are indications that the volume of the keel produces wave-making resistance. This has been known to yacht designers for a long time and they used to 'fair in' the volume of the keel in the Curve of Cross Sectional Area to avoid abrupt changes in the lengthwise distribution of the volume. In the extreme the volume of the keel was taken out of the volume of the hull so as to keep the curve with and without keel identical.

Beukelman and Keuning[3] (1975) found a similar result in their study on the influence of keel sweep back on sailing yacht performance. The model with the keel with zero sweep, i.e. the one which caused an abrupt change in the cross sectional area curve all other parameters such as wetted area and volume etc. being equal, gave in the upright condition the highest resistance by almost 5%!

In 1995 Keuning and Kapsenberg[4] and in 1997 Keuning and Binkhorst[5] published further results on more systematic research on the appendage resistance both in the upright and the heeled condition. In these studies the forces on the keel and rudder have been measured separately from the forces on the hull and the combination keel and hull. The results of these investigations showed clearly the existence of residuary drag on the keel in the upright condition. Depending on the actual geometry of the keel it varied between 2–5% of the overall resistance in particular with increasing speed.

A typical result is shown in Figure 5.9 for a small, high-aspect ratio, low-volume fin underneath an IACC hull. From these measurements it can be seen that there was a speed dependent interaction between the hull and the keel, i.e. at certain speeds the total resistance was smaller than the sum of the individual components and at other speeds it was larger than the sum.

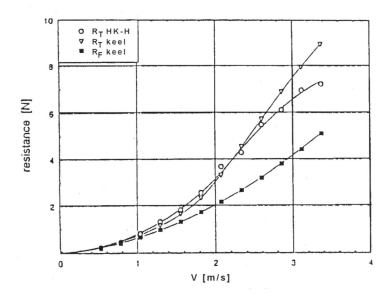

Figure 5.9 *Frictional and Total Resistance of an IACC type keel as measured directly on the keel underneath the hull (keel) and as the differential between the appended hull (HK) and the bare hull resistance (H) i.e. keel = (HK–H).*

5.4 HEELED RESISTANCE

So far only the residuary resistance of the hull in the upright condition has been considered. But sailing yachts are influenced by the forces on the sails, which not only produce the propelling force but also a heeling and a trimming moment. These moments cause the yacht to take on a heeling angle and, if the trimming moment of the sails is not counteracted by another moment for instance by moving the crew weight aft, a change in running trim leading to a more 'bow down' attitude.

Change of viscous resistance due to heel

The angle of heel at which the boat sails will cause the underwater part of the hull to become asymmetrical in shape and, dependent on the shape of the hull, will lead to a change in the wetted area of the hull. This change in wetted area (WSA) with the heeling angle is different for each particular hull under consideration. It can easily be determined for a hull using hydrostatic calculations if a lines plan of the yacht hull is available. Before a lines plan is available an approximation may be obtained from a polynomial expression based on a regression analysis performed on all the hulls of the Delft Systematic Yacht Hull Series.

For a range of heeling angles between 0–30° the equation [5.6] was found to approximate the wetted area of all the models within the DSYHS with great accuracy. The coefficients are given in Table 5.2.

$$S_{c(\varphi)} = S_c \cdot \left\{ 1 + \frac{1}{100} \cdot \left[s_0 + s_1 \cdot \left(\frac{B_{wl}}{T_c}\right)^2 + s_2 \cdot \frac{B_{wl}}{T_c} + s_3 \cdot C_m \right] \right\} \qquad [5.6]$$

S_c = WSA hull upright
$S_c(\varphi)$ = WSA hull at heel angle φ
T_c = Canoe body draft
B_{wl} = Waterline beam
C_m = Cross sectional area coefficient

Table 5.2 Coefficients of wetted surface polynomial

φ	5	10	15	20	25	30	35
s_0	−4.112	−4.522	−3.291	1.850	6.510	12.334	14.648
s_1	−0.027	−0.077	−0.118	−0.109	−0.066	0.024	0.102
s_2	0.054	−0.132	−0.389	−1.200	−2.305	−3.911	−5.182
s_3	6.329	8.738	8.949	5.364	3.443	1.767	3.497

Change of residuary resistance due to heel

Somewhat more significant is the change in residuary resistance of the bare hull due to heel. When the yacht hull heels over this will, apart from the asymmetry of the hull lines, also cause a change in the distribution of the cross sectional areas over the length of the ship. Depending on the geometry of the hull this will lead to some change in the hull shape parameters as used for the definition in the upright case, like the waterline length, the waterline beam, the canoe body depth and the longitudinal position of the centre of buoyancy. In its turn this last change may lead to a change in trim angle also.

The most influential parameters however on this change in resistance are the Beam to Draft Ratio of the hull and the Longitudinal Position of the Centre of Buoyancy.

A typical result of the change in residuary resistance due to heel is shown in Figure 5.10 for two hulls, Sysser model #1, the parent model of Series 1, and Sysser model #28 of Series 3.

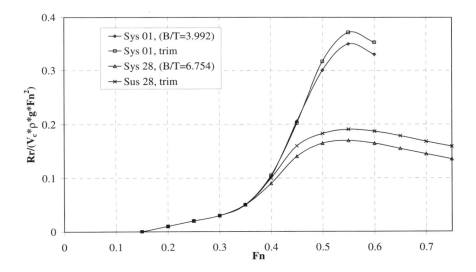

Figure 5.10 *Influence of heel on specific residuary resistance of two different hulls of the DSYHS (R_r = residuary resistance, V_c = volume of displacement).*

The difference in behaviour of the residuary resistance between these boats is obvious from these results. Although relatively speaking at lower speeds the change in residuary resistance might be considerable the effect on total resistance is only significant at speeds above Fn = 0.35. It should be noted that the plots present the specific residuary resistance per ton of displacement divided by the Froude number squared. So at low Froude numbers the differences are exaggerated in the plot.

A similar procedure has been carried out with all the models of the DSYHS in the upright condition but now investigating the influence of the bow down trimming moment, which is imposed on the hull by the driving forces 'high up' on the sails. A typical result is shown in Figure 5.11 once again for two models out of the DSYHS.

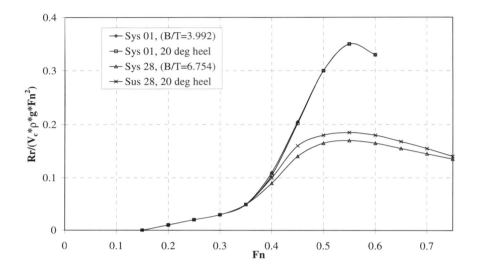

Figure 5.11 *Influence of trim on the specific residuary resistance of two different hulls of the DSYHS.*

From these results it is immediately obvious that this trimming effect has a strong effect on the resistance of the hull. At low speed the effect is small due to the fact that the trimming moment and hence the trim is small. With increasing speed the effect becomes more severe and the more so for the more modern hull forms of Series 2, 3 and 4, even though their relatively lower specific resistance means that they have a smaller trimming moment. In the end at the higher speeds the difference may amount to 10–15% of the residuary resistance.

In actual sailing this brings in a size-related effect to the sailing yacht performance:

Smaller boats with relative larger crew weights compared to their displacement and sail area, may be able to 'counter balance' the negative effect of trim on the resistance by moving the crew aft and thereby neutralising the bow down trimming moment of the sails.

On bigger and heavier boats this will almost certainly not be possible and therefore these boats will be penalised by the negative effect of the bow down trim on the resistance and therefore sail at a lower speed.

5.5 INDUCED RESISTANCE

In the introduction on the various resistance components the nature of the induced resistance has already been explained. Its origination is directly coupled to the lift generation itself and its magnitude is strongly coupled to the amount of downwash generated by the 'wing' and therefore to the efficiency of the wing as a lift generator.

The analogy with the 'aeroplane' wing has proved very useful in understanding and predicting the induced resistance of a sailing yacht. In fact the sideforce production of a sailing yacht may be simplified to the sideforce production of three separate wings working in each others vicinity and influencing each other, i.e. the keel and rudder as wings with moderate to reasonably high aspect ratios on one side and the bare hull as a very low aspect ratio wing on the other side So we will consider in somewhat more detail the effects of wing section and planform on the induced resistance.

The wing geometry and planform definitions

Figure 5.12 shows the commonly used parameters that define wing geometry.

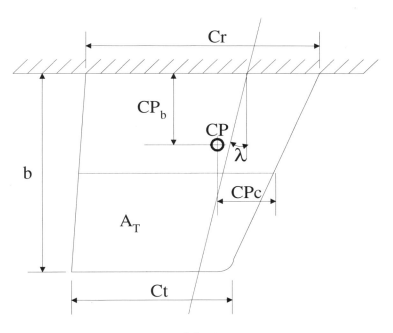

Figure 5.12 *Keel planform geometry definitions.*

$C = \frac{1}{2}(Cr + Ct) =$ mean chord
$Cr =$ root chord

C_t = tip chord
b = span
λ = sweep angle
A_t = wing area
AR = aspect ratio = b^2/A

From theoretical calculations it has been demonstrated that the induced resistance component is minimal for a given wing when the wing has an elliptical load distribution over its span.

In practice there are several methods of achieving elliptical loading for the wings: the planform of the wing can be made elliptical or the section shape (profile) can be adjusted over the span to achieve this loading profile. An exact elliptical planform is not strictly necessary for an elliptical loading. In practice the introduction of a taper ratio in the order of magnitude of $c_t/c_r = 0.6$ is for most practical applications already quite effective in this respect.

Also the use of a sweep back angle (λ) will lead to a change in a spanwise distribution of the lift, i.e. increasing sweep angle leads to a higher loading of the wing tip. This in its turn will lead to an increase of the induced resistance.

Interesting results in this respect are shown by Kerwin *et al.*[6] in 1973 in their paper on Sailing Yacht Keels. Based on lifting surface calculations and model experiments in the water tunnel, they showed in detail the influence of sweep angle and taper ratio on the lift and induced resistance of a yacht keel in the absence of any free surface effects.

The effect of the wing tip

Because of the fact that the induced resistance is strongly related to the strength and shape of the tip vortex, changes to the shape of the tip of the wing may have influence on the induced resistance. The flow around the wing tip, i.e. from the high-pressure side to the low-pressure side, must be 'hindered' as much as possible in order to minimise the induced resistance.

A very good way of doing this is by means of an 'end plate'. This is a large plate at the wing tip which extends well beyond the chord of the tip both upstream and downstream and protrudes above the section profile. (See Figure 5.13.) This end plate very effectively blocks the flow around the wing tip and inhibits the vortex generation. However it also produces considerable additional resistance, in particular when it is not positioned correctly in the 'free' stream. In most cases the presence of a sailing yacht hull above the keel and the rudder may be considered to be an example of such an end plate.

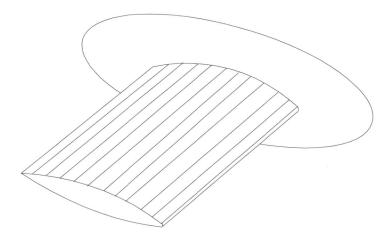

Figure 5.13 *End plate of a wing tip.*

On the other end of the wing various shapes of the wing tip are often used to minimise the tip losses, such as squared and triangular shaped wing tips and also bulbs. From the previously mentioned research from Kerwin *et al.*[6] it became evident that for the induced resistance the V-shaped tip chord gave the least resistance and the round tip the most. In the condition with no lift however the ranking of these wing tips was reversed, i.e. the round tip yielded the least resistance. So a trade-off between upwind (lift and induced resistance on top of 'normal' resistance) and downwind (no lift just 'normal' resistance) sailing has to be made.

As far as bulbs are concerned, these deliver a considerable additional resistance, primarily because of their large wetted area and high 'form drag', and they adversely affect the lift generated and induced resistance. This means that the beneficial effect of the use of a bulb has to be justified completely by the gain in static transverse stability of the yacht through the lower centre of gravity made possible by the weight of the ballasted bulb.

The effect of the aspect ratio of the wing

A very important parameter of the wing planform on the induced resistance characteristics of a wing is the Aspect Ratio, i.e. the ratio between the wing span and the wing area. A long slender wing has a high aspect ratio. A high aspect ratio effectively means that the effect of the wing tip on the overall performance of the wing is small, i.e. the wing behaves more and more as if it were in a two-dimensional flow. The influence of the aspect ratio on the induced drag is clearly demonstrated in Figure 5.14 taken from Theory of Wing Sections by Abbott.[7] Here the drag coefficient C_{DI} is presented as function of the lift coefficient Cl for various aspect ratios of the same wing.

From this figure it is obvious that the largest gains in reducing the induced resistance for a given lift coefficient can be made in the area of the relatively low aspect ratios wings. This is the aspect ratio range in which the majority of the keels of the cruiser–racers are placed and so large gains can be made here by only small improvements on the aspect ratio of the yacht's keel, commonly accomplished by increasing the depth of the keel. At higher aspect ratios this gain becomes smaller.

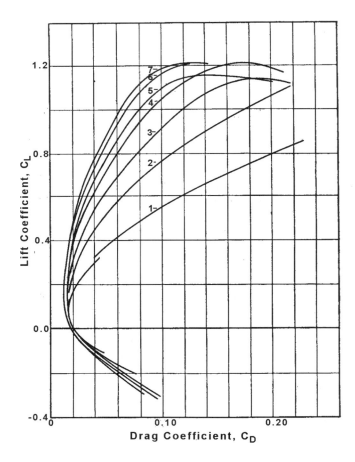

Figure 5.14 *Lift Cl–drag Cd plots for wings with increasing aspect ratio.*

The effect of the free surface

So far only the induced resistance of a wing in a situation in which it is completely surrounded by fluid (i.e. deeply submerged), has been considered. Until 1975 this was normal practice and keels were investigated for their effectiveness in water circulation tunnels or wind tunnels without any free surface.

In 1975 Beukelman and Keuning[8] presented their measured results on a systematic study of the effectiveness of one particular keel with different aspect ratios and sweep angles and varying angles of heel tested on one particular model in the Delft Towing Tank. In this research special attention has been paid to an additional 'induced' resistance effect, which arises due to the fact that the pressure field around the keel is moved towards the free surface due to the heeling angle of the yacht. This presence of the pressure field near the free surface generates waves on this free surface, which in its turn manifests itself as resistance. Since its existence and magnitude is directly coupled to the lift generation itself, this is also an 'induced' resistance component. The waves created by the keel pressure system are clearly visible in Figure 5.15.

When the sweep angle of the keel is increased the pressure field is spread out over a longer portion of the free surface. In addition the loading tends to become more concentrated near the tip of the wing with increasing sweep angle. So with higher sweep the wave generation and so the induced resistance should be reduced when the yacht is heeled. This was fully confirmed by the measurements on the models: the induced drag appeared to be strongly dependent on sweep angle of the keel, generally decreasing with increasing sweep up to angles of 40–45°. Bringing the pressure field away from the free surface obviously improves the efficiency of the keel when the yacht is heeled.

Much later (1983) this led to the development of the 'upside down' keel, i.e. inverse taper ratio and the use of winglets. It is worth noting that some of the findings of the research projects carried out on lift and drag of keels without taking into account the free surface effects may lead to erroneous results. For instance Kerwin found an increase in the induced drag due to an increasing sweep angle, which is only correct without free surface effects being present.

The tendency that the induced drag increases with sweep back angle for a given lift in the upright condition was generally found to be not true under heeled conditions. This is caused by the significant influence of the interference of the keel and hull wave systems. This important interaction between the pressure field around the keel and the free surface may not be neglected in the keel design, and is discussed further in Chapter 7.

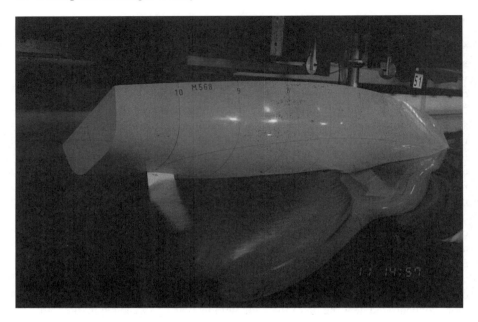

Figure 5.15 *Windward side wave pattern of a hull with keel, leeway = 4°.*

Approximation using the effective keel method

From the results presented in the previous sections discussing the dependence of the induced resistance on various design parameters it became obvious that the aspect ratio is the prime driver for the induced resistance of a keel.

In order to assess the induced resistance of an arbitrary yacht design, i.e. a keel–hull combination, use again has been made of the results of the tank test data with the models in the heeled and yawed condition of the Delft Systematic Yacht Hull Series (DSYHS). All tests with the models of the DSYHS have been carried out with the models fitted with the same keel and rudder. This has been done for the sake of consistency throughout the whole series and makes it possible to determine the effect of the hull on the induced resistance. For a more detailed description of the tests reference is made to the various publications.[9,10,11]

The method used to analyse all these data and to set-up a prediction tool taking into account the prime parameters is 'The Equivalent Keel Method', as described by Gerritsma.[12]

In this method the keel (and the rudder) are extended to the waterplane of the yacht in order to be able to calculate the lift and the induced drag of the wing using the known formulations from aerofoil theories.

An even further analysis of the data revealed that the induced resistance is to a very large extent only dependent on the actual span of the effective wing. So an 'effective draft' (T_e) was formulated to calculate the induced drag of an arbitrary hull. The induced resistance (R_i) is calculated from equation [5.7].

$$R_i = \frac{F_h^2}{\pi T_e^2 q} \qquad [5.7]$$

F_h = heeling force
q = dynamic head ($1/2\ PV^2$)

The assumption that the induced resistance of such a complex wing system as a sailing yacht hull with appendages behaves like a wing so that the quadratic dependency of the induced resistance on sideforce holds true, may be seen from a typical example as presented in Figure 5.16 for two different models of the DSYHS.

Figure 5.16 shows clearly the effect of the free surface on induced resistance. Model 16, with a low canoe body beam/draft ratio has a lower rate of induced resistance increase with heel angle than model 28. The wide shallow hull shape of model 28 causes the keel and rudder root to move more quickly towards the water surface as heel angle increases.

The solid lines in Figure 5.16 are calculated from the regression formula described in reference 10, which uses the DSYHS results to calculate induced resistance based on the hull parameters.

It should be emphasised however that only an approximation capable of dealing with a large variety of yachts may be obtained using this method. Further extensions and refinements of the present method are still being developed, and may yield more discriminative results later, but a very detailed answer on the influence of all the design parameters of the hull and the appendages may only be obtained by detailed calculations combined with extensive tank testing.

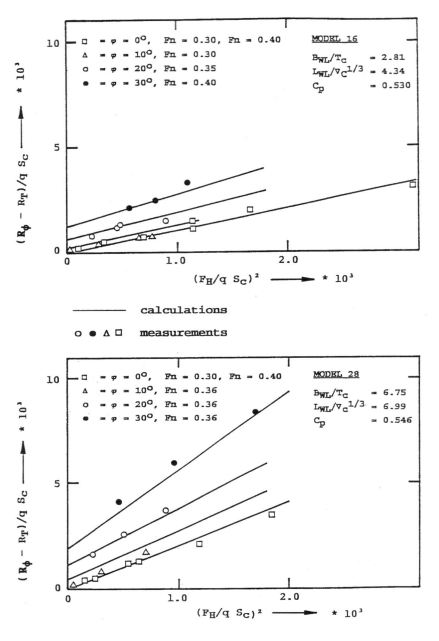

Figure 5.16 *Induced resistance as a function of heel angle and sideforce for two different beam/draft ratio hulls.*

5.6 HYDRODYNAMIC SIDEFORCE

The hydrodynamic forces on a sailing yacht discussed so far were all more or less directed in the direction of the motion of the yacht, i.e. resistance components. Fortunately those are not the only forces generated by the moving sailing yacht hull. The hull and more in particular the appendages generate also considerable sideforce, which is intended to counteract the lateral forces generated by the wind on the sails.

In assessing the physics of the sideforce generation, just as with the induced resistance, the analogy with the wing is evident.

The influence of finite span

In previous sections we have seen how the flow field around a cross section of a wing can be considered to be composed of two separate potential flow patterns: one uniform flow of the incoming stream and a circulating flow (or vortex) 'bound' on the wing span.

The lift increases with increasing circulation for a wing of a given planform and section profile, this occurs with an increasing angle of attack. This is true until an angle of attack is reached at which the flow on the back of the section, the suction side, is no longer capable of 'following' the section profile and so the flow 'separates' from the foil (the foil 'stalls'). The result of this suction side flow separation is a quick deterioration of the lift and an associated sharp increase in the resistance.

This is most clearly demonstrated in Figure 5.17 derived from 'Theory of Wing Sections'[7] showing for one particular section profile the dependency of the lift coefficient on the angle of attack for various aspect ratios.

From this figure it is clear that the high aspect ratio wing is far more effective in generating lift at the same angle of attack than the low aspect ratio wing. It should be noted however that ultimately the maximum attainable lift coefficient of all the foils is the same irrespective of the differences in aspect ratios. Another striking difference between the various wings is the difference in angle at which they start to stall: the high aspect ratio wings generate a high lift at small angles of attack but they stall very soon, the low aspect ratio wings generate much less lift per degree of angle of attack, i.e. the lift curve slope is considerably less, but the angle of attack at which the foil stalls is much higher.

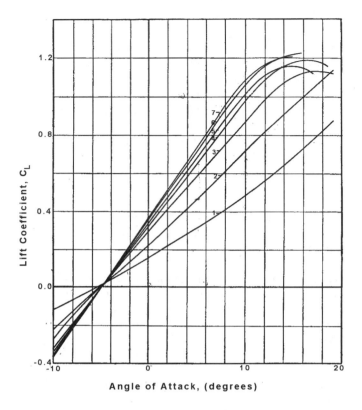

Figure 5.17 *Lift coefficient against angle of attack for a wing with increasing aspect ratio.*

So an effective wing in respect of high lift production at small leeway angles and at the cost of minimal induced drag is found in the high aspect ratio wing. These high aspect ratio foils enable the yacht to sail at relative low leeway angles at least when the speed is sufficiently high. The keel becomes effective also because a reduction in the wetted area of the keel leads to lower frictional resistance. Obviously environmental conditions such as water depth as well as constructional considerations may lead to restraints on the keel span.

Coming out of a tack with these high aspect ratio keels may cause difficulties if the boat is not properly handled, because just after tacking the lift coefficient of the keel is high. This leads to high leeway angles and the danger of 'stalling' the keel, due to the yacht's low speed.

Another drawback of these high aspect ratio keels in this respect is found in the behaviour in waves when sailing 'upwind'. Due to the combined effect of the ship motions and the waves the angle of attack of the keel is changing continuously and rapidly. In addition to this the added resistance of the yacht when sailing in waves may cause lower forward speeds to be maintained in those conditions. This results in a relatively higher loading of the keel and a strongly varying angle of attack. This may lead to stalling of the keel in those conditions.

The application of a very high aspect ratio rudder introduces similar problems and calls for very careful application of rudder angle.

The influence of the section profile

The lift per unit of planform area of a wing with infinite span may be considered to be dependent only on the speed of the approaching fluid, the angle of attack and the shape of the section. The latter two determine the amount of circulation around the section or the strength of the bound vortex.

In general it may be stated that the important parameters describing the sectional cross section shape which determine the pressure distribution on the section (and the circulation at a certain angle of attack) are :

1. Whether the cross section is symmetrical or asymmetrical
2. The thickness ratio of the section, i.e. the maximum thickness of the section divided by the chord length
3. The longitudinal position along the chord length of the maximum thickness of the section

Considering the symmetric versus asymmetric shape it should be noted that in sailing yacht applications, in general, the shape of the sections is symmetrical because the 'wings' (i.e. keel and rudder) on a sailing yacht, in contrast to aeroplane applications, have to deliver equal performance on both tacks. Asymmetry in the section shape however may be introduced by the application of flaps at the trailing end of the section. The general effect of the application of such a trailing-edge flap is an increase of lift at a given angle of attack when compared to the section without flap, however at the cost of a slightly higher drag.

The amount of thickness (i.e. t/c) in general increases the maximum lift attainable but at the cost of a slightly higher resistance. The thicker foils are also less sensitive to stall than the slender foils.

The longitudinal position of the maximum thickness of the aerofoil section and the detail section design influence the drag coefficient because they determine the extent of laminar flow over the forward part of the aerofoil. By moving the position of maximum thickness aft and increasing the thickness to chord ratio pressure distributions that promote favourable conditions for the laminar boundary layer can be achieved. By moving the transition point aft on the aerofoil the drag coefficient is reduced. However, if the maximum thickness is taken too far aft the aerofoil becomes prone to boundary layer separation at relatively low lift coefficients. Thus in designing a section for a keel or rudder a balance must be struck between achieving low drag at low lift coefficients and potentially higher drag at high lift coefficients. Striking a correct balance is difficult because, as discussed earlier, the actual lift coefficient will fluctuate depending on the point of sailing and the wind conditions. Additionally the effects of the ambient conditions such as surface roughness, waves and turbulence in the water all affect the degree to which the laminar boundary layer can be preserved.

To find an optimal shape of the cross section (given a set of desired properties of the wing) a tremendous amount of research has been carried out on all kinds of different section shapes. Outstanding in this respect is the work carried out through NACA (USA) of which much data is compiled in the book *Theory of Wing Sections*,[7] and similar work by Wartmann.[15]

The aim of these studies is to find a pressure distribution over the foil which is optimal for the expected use of the wing in question. For a sailing yacht keel in general this will be focused on a minimisation of the resistance and a maximum lift to drag ratio.

The influence of the free surface

Just as is the case with the induced resistance of a wing there is also an influence of the free surface on the lift generating properties of a wing. This is represented in the 'equivalent keel' approach by mirroring the extended keel around the waterline. By doing so the effective aspect ratio of the equivalent keel is more than doubled when compared to the geometrical aspect ratio of the actual keel. As soon as the yacht gets heeled however the pressure distribution over the keel is influenced by the presence of the free surface.

This effect is also influenced by the characteristics of the hull above the keel, i.e. a wide and shallow hull will tend to move the keel closer to the free surface when heeled than a narrow and deep hull. The latter however will also have less of an 'end plate' effect on the keel.

One other specific problem with the wings in the close proximity of the free surface is 'ventilation' of the foil. When the wing generates lift this is achieved by creating a high pressure at one side and a low pressure on the other side of the wing sections. Dependent on the specific section profile selected in general there exists a considerable peak in the pressure distribution at the low pressure side. When the pressure in this peak reduces further and further with increasing loading on the foil, and the top of the foil is on or close to the free surface, air can be sucked in and this will result in a sudden loss of the low pressure peak and therefore of the lift.

Most vulnerable to this ventilation phenomenon is the rudder which under heel may even penetrate the free surface (Figure 5.15) and which may be subjected to high angles of attack (and loading) under conditions where increased rudder action is asked for with respect to 'balancing' the yacht. In addition the low pressure side is by definition always the closest to the free surface.

The effects of this ventilation phenomenon may only be reduced by:

1. Eliminating heel as much as possible
2. Keeping the rudder submerged as much as possible
3. Selecting a section profile with a decreased low pressure peak in the pressure distribution.

Approximation of the sideforce as function of heel and leeway

To be able to approximate the sideforce of an arbitrary keel under an arbitrary hull once again use has been made of the results obtained from the tests with the models of the Delft Systematic Yacht Hull Series.

The emphasis in this approach is on the effect of the hull geometry on the lift producing capabilities of the keel and the effect of heel angle. Initially, all the models of the DSYHS had a similar keel fitted for the heeled and yawed tests. In addition to these tests however extra tests have been carried out on two models with keels fitted with a different span, 0.25, 0.50 and 1.20 times the span of the original keel. For details see reference 11 where there is given an empirical formulae for the relation between heel, leeway and sideforce.

A typical fit of the expression to the experimental data is presented in the Figures 5.18 and 5.19 for two completely different models of the DSYHS, i.e. the parent model of Series 1 with the half-span keel and for a very high beam to draft ratio model of Series 2. As may be seen the match of the approximation with the experimental data is very good.

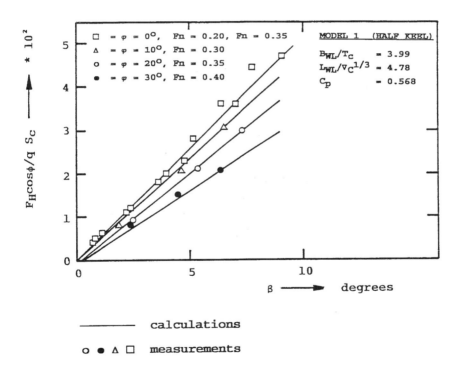

Figure 5.18 *Sideforce as a function of leeway for various heeling angles. (Model 1 half-draft keel)*

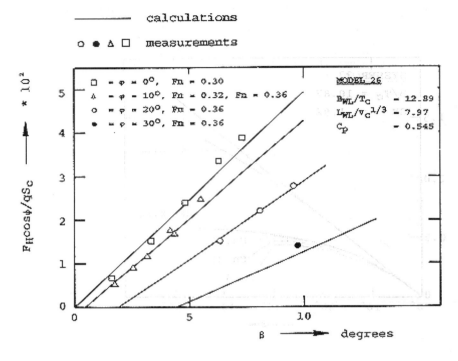

Figure 5.19 *Sideforce as a function of leeway for various heeling angles. (Model 26)*

REFERENCES

1. Principles of Naval Architecture, SNAME.
2. Marchaj, C.A., Aero-Hydrodynamics of Sailing, 1988.
3. Beukelman, W. and J.A. Keuning, The Influence of Fin Keel Sweep-back on the Performance of Sailing Yachts, Delft University of Technology, Shiphydromechanic Laboratory,1975.
4. Keuning, J.A. and Kapsenberg, G., Wing - body interaction on a sailing yacht, Report 1019-P, 1995.
5. Keuning, J.A. and Binkhorst, B.J., Appendage resistance of sailing yacht hull, 13th Chesapeake Sailing Yacht Symposium, 1997.
6. Kerwin, Justin E. and Halsey C. Herreshoff, Sailing Yacht Keels, Massachusetts Institute of Technology, Cambridge, 1973.
7. Abbott, Ira H. and Albert E. von Doenhoff, Theory of wing sections, 1959.
8. Gerritsma, J., Coarse Keeping Qualities and Motions in Waves of a Sailing Yacht, 3rd AIAA Symposium on Aero-Hydrodynamics of Sailing, California, 1971.
9. Gerritsma, J., Onnink, R. and Versluis, A., Geometry, resistance and stability of the Delft Systematic Yacht Hull Series, 7th HISWA Symposium, Amsterdam, 1981.

10. Gerritsma, J., Keuning, J.A. and Versluis, A., Sailing yacht performance in calm water and waves, 11th Chesapeake Sailing Yacht Symposium, SNAME, 1993.
11. Keuning, J.A. and Sonnenberg, U.B., Developments in the Velocity Prediction based on the Delft Systematic Yacht Hull Series, RINA International Conference on the Modern Yacht, 1998.
12. Gerritsma, J., Glansdorp, C.C. and Moeyes, G., Still Water, Seakeeping and Steering Performance of 'Colombia' and 'Valiant', Delft University of Technology, Shiphydromechanic Laboratory, 1974.
13. Hoerner, S.F., Fluid Dynamic Drag, 1965.
14. Hoerner, S.F., Fluid Dynamic Lift, 1975.
15. Althaus, D. and Wartmann, F.X., Stuttgarter Profilkatalogl Fried. Vleweg and Sohn. 1979.

CHAPTER 6

DYNAMIC BEHAVIOUR OF SAILING YACHTS IN WAVES

J.A. Keuning
Delft University of Technology

6.1 INTRODUCTION

So far the stationary sailing condition of a sailing yacht at an arbitrary heading with respect to the true wind has been dealt with. It is known however that in reality the conditions in which a sailing yacht sails are most of the time significantly affected by the presence of wind generated surface waves. The simplification of assuming no waves to be present is not always justifiable, in particular in optimisation procedures, since it may not be assumed that the yacht that performs best in calm water conditions when compared with other designs also performs best in a seaway.

These waves generate forces and moments on the sailing yacht hull and provoke motions in all six degrees of freedom. These wave forces and the resultant motions will be shown to be functions of the waves encountered, the geometry of the hull and the forward speed of the yacht. In particular the pitch, heave and roll motion cause all kinds of unsteady effects, for instance on the lift generating capabilities of the sails and appendages.

A particular phenomenon of great interest is an additional resistance component (RAW) experienced by the yacht due to the presence of these surface waves. This added resistance manifests itself most severely in the upwind sailing condition and may reduce the upwind performance of a boat to a considerable extent. In following waves an opposite phenomenon may arise; the yacht may be 'propelled' by the exciting forces of the following waves. High forward speeds become possible when the yacht is on the front face of the wave. This is known as 'surfing'.

6.2 ENVIRONMENTAL CONDITIONS

In order to be able to relate the behaviour of a yacht with respect to the surface waves it is essential to quantify the physical properties of these surface waves. It is customary to simplify the rather complicated and very irregular 'shape' of the water surface as we all know it, i.e. short-crested waves of various height following each other, by a system that is formed by the superposition of a series of regular harmonic waves each a different wave length and a different wave height.

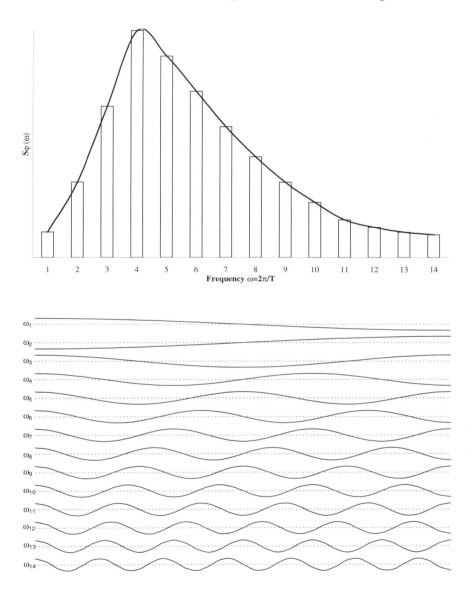

Figure 6.1 *Combination of wave frequencies to create a wave spectrum.*

It is known from mathematics that the summation in the time domain of only a few harmonic 'components' each with its own length and height yields a rather irregular signal, so the reverse procedure in which a large number of different components is used to 'unravel' an irregular wave height signal is generally considered to be valid at least in the linear theory.

The normal method of describing a particular seastate is by means of a 'wave energy spectrum', in which a particular 'measure' of the wave heights of the various components forming the irregular seastate are presented on a basis of their frequency. A typical example of such a wave spectrum and the way it is formed by the various components is presented in Figure 6.1.

Each seastate may now be characterised by its spectrum. In addition each spectrum may be characterised by a number of parameters of which only the most relevant will be presented here. These are:

- The area underneath the spectrum and the frequency (horizontal) axis M_o.
- The significant waveheight ($H_{1/3}$.) This is the average of the one third highest waves in the given seastate, $H_{1/3} = 4 \sqrt{(m_o)}$. The significant waveheight closely correlates with the estimate of an experienced observer of the average waveheight in a seastate.
- The peak period Tp of the spectrum, which is the wave period belonging to the component with the highest waveheight in a particular spectrum.

It should be realised that there exists a certain relationship between the wind strength, the fetch of the wind (i.e. the over-water distance over which the wind is blowing), the duration of the prevailing wind speed and the spectrum of the wind waves for any particular area. For the North Atlantic this relationship is presented in Figure 6.2.

Since the 'irregular' seastate may be considered to be a superposition of a large number of 'regular' harmonic waves, of which all the physical properties can be easily described by mathematical formulas, and the ship's behaviour in waves as being a linear system, the motions of a ship in any 'realistic' irregular seaway may be computed and analysed by studying the response of a ship to harmonic waves. This simplifies the computations enormously.

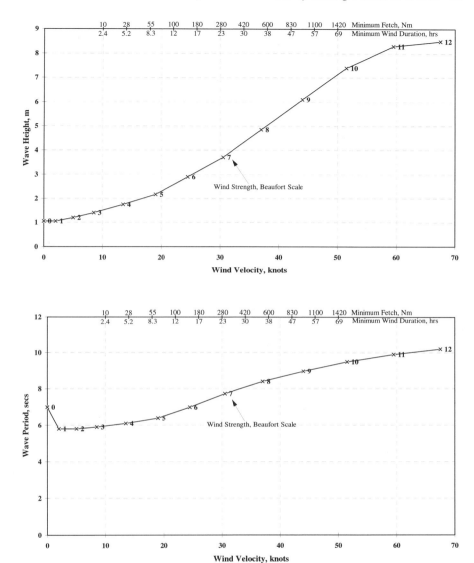

Figure 6.2 *Significant wave height and average period of fully developed seas.*

6.3 MOTIONS IN WAVES

To calculate the motions of a ship in irregular waves the response of the ship to regular harmonic waves must be known. Since the ship is considered to be a linear system this response is given in a nondimensional form as a response amplitude operator (RAO) and a phase lag between the wave (excitation) and the motion (response) for a large number of harmonic waves with different wave length. The six motions of interest are defined in Figure 6.3.

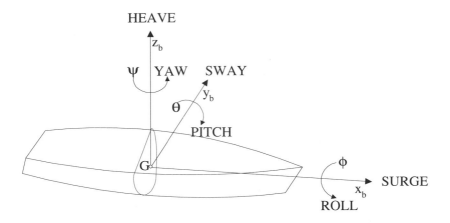

Figure 6.3 *Axis system for ship motions.*

The RAO's for the six degrees of motion, i.e. (surge X, sway Y, heave Z, roll ϕ, pitch θ and yaw ψ) represent a nondimensional response on a basis of wavelength or frequency. The translations X, Y and Z are divided by the wave amplitude ζ and the rotations ϕ, θ and ψ are divided by the waveslope ($2\pi\, \zeta/\lambda$).

Typical heave and pitch RAOs are shown in Figure 6.5 for a yacht in head waves at a Fn of 0.35. The general shape of these RAOs shows practically no response (i.e. RAO = 0) in the very short waves, some resonance (i.e. RAO > 1.0) at a wavelength between 1.0 and 1.5 times the ship length and a response equal to the waveheight or wave slope (i.e. RAO = 1.0) in very long waves, in which the ship may be considered to follow the wave contour exactly. The natural periods of heave and pitch are very important for the behaviour of a yacht in waves. If one of these natural motion periods is equal to the period of the excitation forces, i.e. to the period of wave encounter, violent motions may result, i.e. RAO >> 1.0.

Finally to compute the response of the yacht in a given seaway, characterised by a given wave spectrum, the spectral density of this wave spectrum given on a basis of the wave frequency, is multiplied by the RAO squared for the particular motion under consideration, which has to be given on the same basis. Doing this for the entire frequency range of the wave spectrum yields the energy spectrum of the motion. From this motion spectrum the statistical values of interest may be obtained, such as the significant amplitude etc. This procedure is visualised in Figure 6.4.

An important effect becomes immediately obvious from this: only where the wave spectrum and the RAO 'overlap' is wave energy transferred from the waves into the motion. So avoiding high values of the significant values of the motion under consideration in the given wave spectrum may be achieved by the following situations and/or actions:

- Minimise the values of the RAO over the entire frequency range of the wave spectrum
- Avoid high values of the RAO (i.e. resonance) in particular in the frequency domain of the spectrum under consideration but in particular at or near the peak period of the given wave spectrum
- If the RAO and the wave spectrum are shifted in the frequency range with respect to each other in such a way that they do not or hardly 'overlap' the response of the yacht is minimised

A few of these possibilities are within the scope of actions of the designer and the user of the yacht.

It is of great interest to the designer to know how the various hull parameters influence the motions. To illustrate this the results of calculations will be shown carried out with a systematic variation of one particular hull shape. The hull shape chosen is the parent of DSYHS Series 4, which is a representative hull shape for current designs. From this model new models have been derived changing the Length to Beam Ratio (L/B), the Length–Displacement Ratio (L^3/∇) and the Pitch Radius of Gyration (Kyy/L). The influence of this on the heave and pitch motions in head waves is presented in Figures 6.5, 6.6 and 6.7.

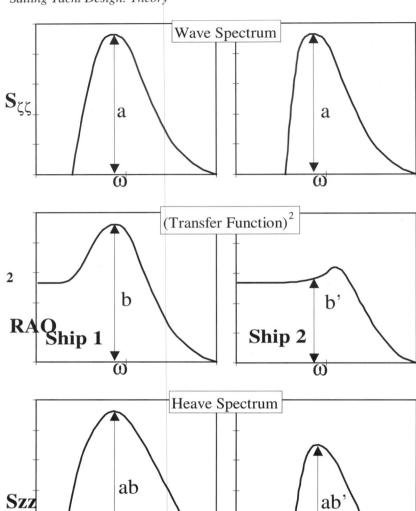

Figure 6.4 *Combination of wave spectrum with RAO to determine motion spectra for two different yachts.*

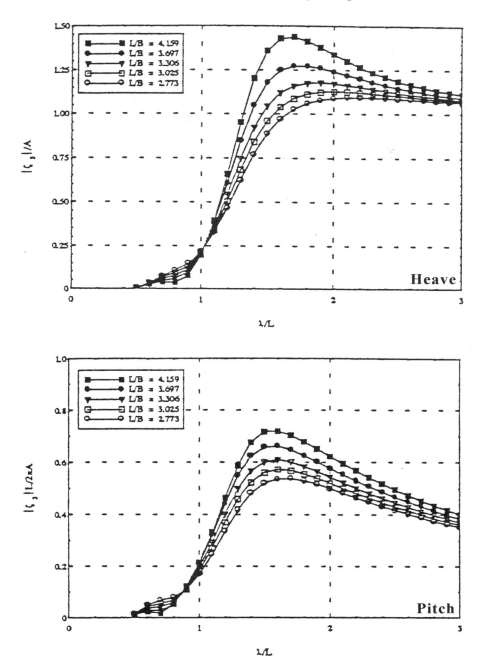

Figure 6.5 *Influence of L/B Variation on the Heave and Pitch RAOs.*

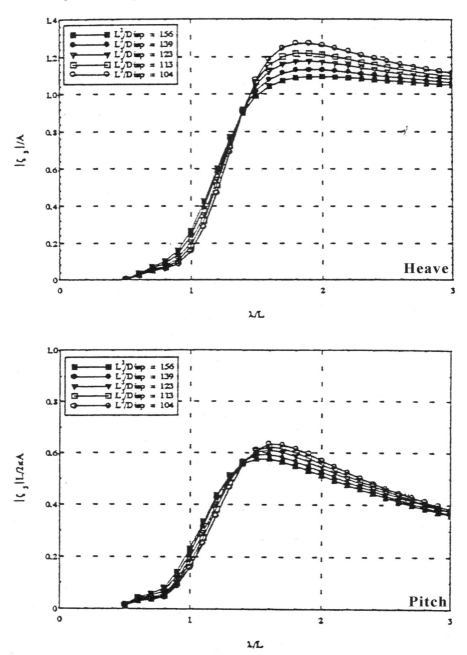

Figure 6.6 *Influence of length/displacement ratio on Heave and Pitch RAOs.*

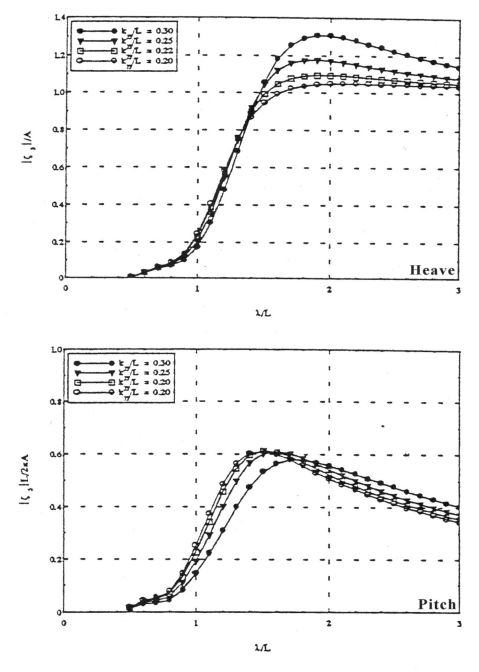

Figure 6.7 *Influence of radius of gyration (Kyy) on Heave and Pitch RAOs.*

It should be mentioned that the results presented refer to head waves and one particular speed corresponding to Fn = 0.325 (typical upwind condition) only. The results are plotted on a basis of the nondimensional parameter 'relative wavelength', i.e. the ratio between the wave length and the waterline length of the ship. It should be realised when analysing these figures that for the bulk of the 'fleet', i.e. yachts with waterline lengths between 7–15 m, the waves of interest are generally 'long' with respect to the ship length so the λ/L area of particular interest is where $\lambda/L > 1.0$ and this needs the most attention.

From the presented results it may be concluded that an increase of the L/B ratio (narrower hull) increases the heave and the pitch motions in the range of wavelength longer than the shiplength, i.e. $1.0 < \lambda/L < 3.0$ and for the range in which the wave length is shorter than the shiplength, i.e. $\lambda/L < 1.0$, the reverse holds true. In general beamy hulls have a large damping in the vertical motions.

For the influence of the Length Displacement Ratio similar tendencies may be observed: for wavelength to shiplength ratio's smaller than 1.3 the lighter ship has the higher heave and pitch motions and for longer waves the heaviest yacht has the largest motions.

The effect of changes in the longitudinal radius of gyration show a significant increase in the heave motion in waves longer than 1.3 times the shiplength and somewhat less in the pitch motion with increasing radius of gyration. For the range of shorter waves the difference in the heave motion is small and in the pitch motion much more pronounced. These results for variation of Kyy are obtained from calculations on the base hull of the series only, i.e. for a L/B of 3.31 and a L^3/Displacement of 123.

The trends found by this investigation are found to be in close correspondence with results obtained from tests carried out much earlier with variations of the parent model of DSYHS Series 1, a Standfast 43 from 1970, with a very different hull shape. This justifies the conclusion that the trends are valid for a large variety of yacht hull shapes, at least in a qualitative sense.

Shipmotions and accelerations, particularly, in a seaway are to be minimised as much as possible when the seakindliness of a hull is being considered. Shipmotions may provoke extreme fatigue and even seasickness among those on board, they hamper their freedom to move around the ship and they can even make the performance of certain tasks or handling of the yacht very difficult.

The motions of the ship also have an adverse effect on some of the hydrodynamic and aerodynamic properties of the yacht. The lift generating characteristics may be strongly influenced by the continuously varying angle of attack of the sails and the appendages due to the ever changing induced velocities, which also vary over the span of the sails and appendages in conjunction with their distance to the centre of gravity of the ship.

In addition severe accelerations due to these shipmotions can put great strains on the rigging but also on the structure of the boat, in particular when the hull is moving through large distances relative to the water surface. A well known example of this is the occurrence of slamming in the bow area of the ship.

6.4 ADDED RESISTANCE IN WAVES

If a yacht is sailing in head waves a large increase in the resistance of the yacht is noticed and a corresponding loss of speed occurs. This extra resistance in waves is called 'added resistance' (RAW). The phenomena may be explained as follows:

When a yacht is sailing in bow or head waves the pitching and the heaving motions of the yacht generate damping waves which are superimposed on the incident wave system. These generated damping waves carry energy away from the moving yacht which has to be 'delivered' by the yacht itself. This is experienced by the yacht as an additional resistance. This added resistance can now be found by equalising the work done by the resistance force and the radiated damping waves energy.

A practical method to calculate this added resistance is formulated by Gerritsma and Beukelman[1] and is associated with the so-called 2D linear strip theory approach to ship motions, although more complicated full 3-D theories are also available nowadays. In their approach Gerritsma and Beukelman equated the added resistance experienced by the ship to the summation over one complete wave period of the sum of all local damping waves generated by a number of sections along the shiplength.

The energy of the damping wave was equated to the local relative vertical velocity of the section with respect to the surrounding water. This vertical relative velocity is the vectorial summation of the heave motion, the vertical motion at each point along the hull due to pitching and the vertical motion of the incident wave at that place. This enables the added resistance to be calculated for any given yacht hull sailing in regular harmonic incident waves and a Response Amplitude Operator similar to that for motions may be obtained.

In regular sinusoidal waves the mean added resistance (R_{AW}) may be calculated from equation [5.1]. which is based on the strip theory approach.[2]

$$R_{AW} = \frac{1}{\lambda} \cdot \int_0^{Lwl} \int_0^{T_e} b'V_Z^2 dx_b dt. \qquad [6.1]$$

λ = wavelength
t = time
b' = cross sectional damping coefficient, corrected for the forward speed
V_z = relative vertical velocity of the considered cross section with respect to the water
T_e = period of wave encounter

In irregular waves for a known wave spectrum the mean value of the added resistance may be calculated from the formula:

$$R_{AW} = 2\int_0^\infty \frac{R_{AW}}{\xi a^2} \cdot S_\xi(\omega_e) d\omega_e \qquad [6.2]$$

From the previous explanation it is obvious that the peak in this added resistance curve RAO for a given yacht may be found near the resonant peaks for heaving or pitching of the yacht, because relative velocities are the largest there. In these conditions the immersion of the bow of the yacht is large due to the unfavourable phase of the bow motion with respect to the wave. This is why these natural periods in pitch and heave play such an important role in assessing the added resistance of a yacht. It should be noted that in these resonance conditions the absolute motions are not necessarily a maximum. In very long waves the motion amplitudes can be very large because the yacht more or less follows the waves contours. However the relative motion of the yacht with respect to the wave is very small and so is the added resistance. It should also be noted that for the assessment of the natural periods in heave and pitch of any yacht the total mass and mass moment of inertia in pitch have to be known. This includes the real mass of the yacht as well as the mass of the water 'entrapped' by the accelerating yacht hull known as the 'hydrodynamic added mass' and the 'hydrodynamic added mass moment of inertia'. These quantities may only be obtained by calculation.

To demonstrate this the tests and calculations carried out in 1973 with a Standfast 43 by Gerritsma et al[3] will be summarised here.

Three variations of one particular design were investigated for their behaviour in waves and their upwind performance. The primary change in parameters between the three different designs was the Length Displacement Ratio, which varied from 4.54 (III) to 4.77 (II) to 5.07 (I). For list of particulars of the models see Table 6.1.

Table 6.1 Principal dimensions of models used for seakeeping study

		I	II	III
Length of design waterlines L_{wl}	m	10.00	10.00	10.00
maximum breadth B	m	3.66	3.66	3.66
draught	m	2.15	2.15	2.15
displacement	kg	8207	9759	11443
displacement of hull	kg	7680	9211	10670
centre of buoancy aft ½L_{wl}	m	0.26	0.26	0.34
centre of gravity below DWL	m	0.25	0.39	0.52
prismatic coefficient of hull		0.566	0.572	0.566
effective sail area	m^2	66	71	75
length displacement ratio $L_{wl}/\Delta_H^{1/3}$		5.07	4.77	4.54

The radius of gyration of the actual yacht was measured at full scale and found to be 25% of the overall length of the yacht. It is often assumed that this value may be used for a large number of other yachts. The natural period in pitch was found to be 2.4 seconds (by measurement) and in heave 2.2 seconds (by calculation). It is of interest to note that the added moment of inertia for the yacht was 69% of the real inertia and the added mass in heave is 185% of the mass of displacement. In comparison with merchant ships these values are rather high as was the damping in heave and pitch also, which was attributed to the relatively high beam to draft ratio.

In the investigation three different values for the radius of gyration (Kyy)in pitch were chosen, i.e. the measured one (25% L_{OA}), one higher (27% L_{OA}) and one lower (23% L_{OA}) for each of the design variations. For the three designs the motions and the added resistance in waves were calculated for each of the three radii of gyration. The irregular waves used in this analysis correspond to the spectral density formulation distribution as given by Pierson Moskowitz but no fixed relation between the spectrum and the wind speed is used, so the yachts sail at each wind speed in an independently chosen seastate. In Figure 6.8 the ocean wave spectra used in the calculations are shown together with the added wave resistance operators for the three different radii of gyration as calculated for design (I) with a forward speed of 6.7 knots.

Multiplication of the wave spectral densities with the corresponding added wave resistance operators results in the three curves in Figure 6.9, in which the result of only one wave spectrum is shown as an example. The area under these curves is proportional to the added resistance of the designs in the irregular waves. Figure 6.9 shows an important reason for these differences in the added resistance which lies in the shift of the added resistance operators to lower wave frequencies (longer waves), because in this region of longer wave lengths the wave heights and thus the wave spectral densities increase. It is clearly shown that larger displacements and weights distributed more towards the ends of the yacht both result in a higher added resistance in waves.

92 Sailing Yacht Design: Theory

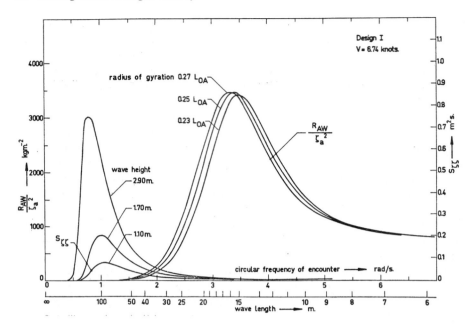

Figure 6.8 *Wave spectra and R_{AW}/ζ_a^2 for Design 1 with changing radius of gyration.*

The favourable effect of the large radius of gyration in the small wave lengths should be noticed, but it is rather small and does not counterbalance the former effect.

When these results on added resistance are applied to the Velocity Prediction Program the results for the optimum speed made good to windward came out as presented in Figure 6.10. The lightest yacht obviously has the highest speed made good in all wind and wave conditions. Similar plots can be made of the effect of the radius of gyration.

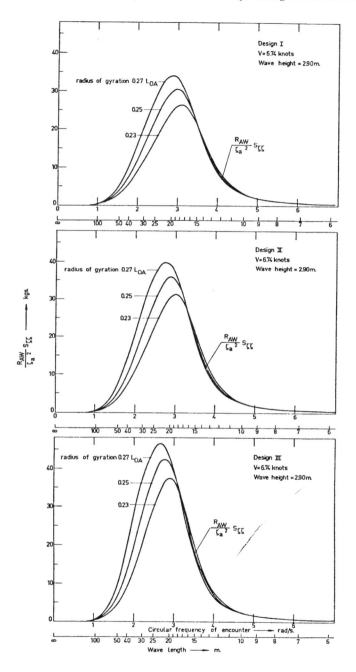

Figure 6.9 *Added resistance responses for three designs with varying radius of gyration in a single wave spectrum.*

94 Sailing Yacht Design: Theory

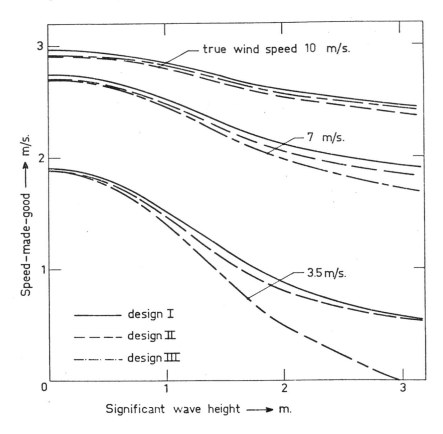

Figure 6.10 *Variation of Vmg with significant wave height at three true wind speeds.*

Using the similar series of systematic models as shown previously for the comparison of the heave and the pitch motions, also the effect of L/B, L/DISP and Kyy on a more modern yacht hull can be demonstrated.

The results of this investigation are shown in Figure 6.11. Clearly also for these hulls the effect of the L/B ratio on the added resistance in waves is the smallest, followed by the Length Displacement ratio and finally the strongest influence is that of the radius of gyration, in particular for the longer waves where the effects is most strongly felt in a typical sea spectrum.

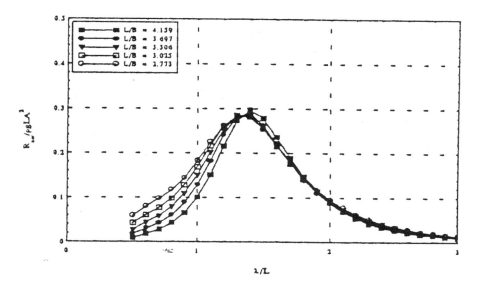

Figure 6.11a *Effect added resistance RAO of changes in L/B ratio.*

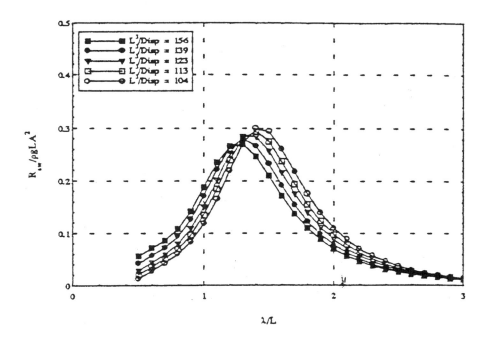

Figure 6.11b *Effect added resistance RAO of changes in K_{yy}/L.*

96 *Sailing Yacht Design: Theory*

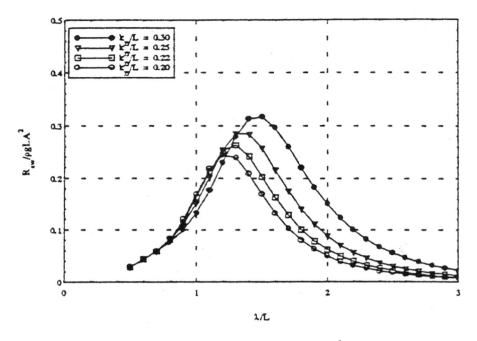

Figure 6.11c *Effect added resistance RAO of changes in $L^3/Disp$.*

The assumptions underlying most of these calculations of the effect of the added resistance in the VPP's is that the influence of the heeling angle of the yacht and the sideforce on the appendages may be neglected. This was investigated by Gerritsma *et al*[4]. by testing two models of the DSYHS, one light and reasonably beamy (resembling a modern IMS racer) and one heavy and very narrow (resembling a 12 m hull), with their appendages in regular head waves. The test were carried out both in the upright condition and at a heel angle of 20°. In addition all the tests were carried out with (5°) and without leeway (and thus sideforce on the appendages). The forward speed corresponded to a typical upwind sailing condition. The heave, the pitch and the added resistance were measured in a series of 10 regular waves tests. The results of the measurements are shown in Figure 6.12.

Dynamic Behaviour of Sailing Yachts in Waves 97

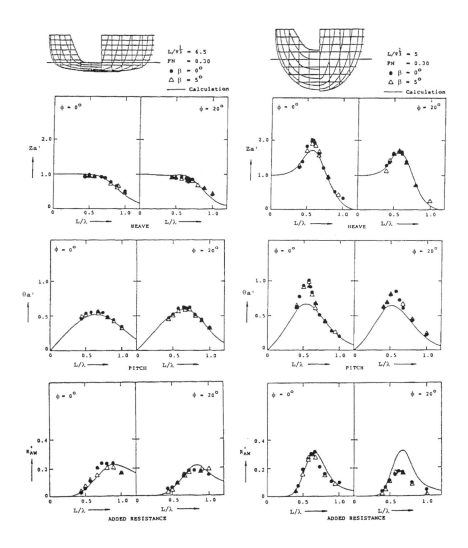

Figure 6.12 *Measured and calculated motion and RAW responses for two hull types at 0° and 20° heel.*

From these results it was concluded that for the light and beamy (high damping) boat the influence of the heeling angle on the motions and added resistance is marginal. Only for the extremely narrow and heavy boat is there a significant difference between the heeled and the unheeled condition.

The influence of the leeway and thus the sideforce on the appendages is not measurable. Therefore omitting these influences on the heave and pitch motion and the added resistance is a quite justifiable simplification. The supposed damping arising from these appendages is obviously of no significance.

Based on a large series of numerical computations performed on a subseries of models of the Delft Systematic Yacht Hull Series and using the above mentioned assumptions of no heel and no appendages, an easy approximation method for the added resistance of an arbitrary yacht has been formulated by Gerritsma et al.[5] using only those parameters which were found to have the most significant influence on the added resistance in waves. The calculations have been carried out for wave headings between 135° (bow quartering seas) and 90° (beam seas) for a variety of forward speeds between Fn = 0.15 and 0.60 and for a variety of wave spectra with peak periods ranging from 2–6 seconds with 0.5 second intervals. A systematic analysis of these results showed that for a constant wave direction, wave height wave period and forward speed the added resistance of all the yachts calculated depends primarily on the parameter:

$$\left(\frac{\Delta^{1/3}}{L_{WL}}\right) \cdot \left(\frac{K_{yy}}{L_{WL}}\right) \qquad [6.3]$$

These results were further elaborated to yield a fully nondimensional approximation to be used for arbitrary yachts. The total results of the added resistance calculations could be summarised by:

$$\frac{100 \cdot R_{AW}}{\rho g L_{WL} H_{1/3}^2} = a \left[\left(\frac{100 \cdot \Delta^{1/3}}{L}\right) \cdot \left(\frac{K_{yy}}{L_{WL}}\right)\right]^b \qquad [6.4]$$

for a nondimensional peak period of the spectrum given by:

$$\frac{T_p}{\sqrt{g/L_{WL}}} = T_1 \qquad [6.5]$$

In equation [6.4], a and b are coefficients of the polynomial expression determined by using a regression fit through the generated data, as described in reference 5.

So now for an arbitrary yacht the added resistance due to waves may be approximated without the necessity of a complex ship motions calculation. The method however does not take all the parameters into account which have an influence on the added resistance and the motions. If more accuracy is wanted use will have to be made of the specific ship motion calculation routines.

6.5 ADDED RESISTANCE IN FOLLOWING WAVES AND SURFING

It should be noted that the results and the analyses presented so far are restricted to waves forward of the beam. For waves aft of the beam the calculations for the added resistance in waves are generally less reliable. In addition the added resistance of the yacht due to waves from those directions is generally small. In fact conditions may occur where the ship is 'propelled' by the presence of larger following waves and so 'surfing' may occur.

This is the ability of a yacht to be 'captured' by a following wave and to accelerate to a speed equal to the phase velocity of that wave. The phenomenon is largely dominated by the relative magnitude of the horizontal wave exciting forces in these conditions, in particular when the yacht is sailing on the front of the wave near the crest there arises a relative large horizontal surge force in the positive direction. The magnitude of this surge force in relation to the yacht's resistance, and the slope of its resistance curve in the speed region under consideration, as well as the actual mass of the yacht and the 'thrust' characteristics of the yacht (sail area etc.) seem to determine when surfing will occur. When the yacht starts to surf high speeds become possible, generally much higher than possible in calm water conditions, speeds which in general are restricted by the phase velocity of the wave itself, because when the yacht 'leaves' its favourable position with respect to the wave the large horizontal surge force in the direction of motion of the yacht will vanish.

To illustrate this behaviour in following waves some of the results of an investigation by Keuning et al.[6] will be presented here. It concerns about the behaviour of three widely different yacht hulls of the DSYHS in (large) following waves. The main particulars of the models are presented in Table 6.2

Table 6.2 Parameters of DSYHS models used in surfing study

Model	#27	#38	#39
Loa (m)	2.31	2.35	2.31
Displacement (kg)	63.4	19.1	18.9
L/B	4.5	3.0	5.0
B/T	2.46	19.3	6.96
$L/V^{1/3}$	5.02	7.49	7.50

First the added resistance has been measured and calculated in the ususal way as described above. From these results it became evident that this 'classical' harmonic approach yielded only small integrated values of the added resistance and in addition to this also the correlation between the predicted and the measured values turned out to be rather poor.

From the measurements only the maximum positive amplitude of the surge force will be presented here although it should be realised that the surge force is strongly dpendent on the instantaneous position of the yacht with respect to the wave. The results for various values of the shiplength to wavelength ratio and two different wave steepness are presented for the three models in Figure 6.13.

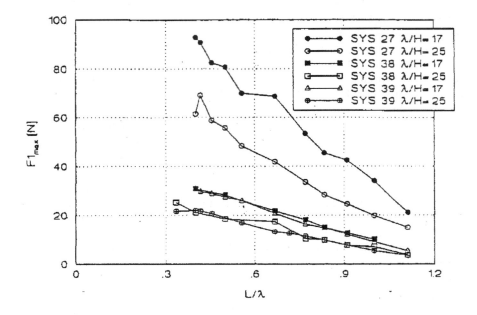

Figure 6.13 *Maximum surge force in following regular waves on three models as a function of wavelength.*

From Figure 6.13 it may be seen that the surge force is roughly proportional to the displacement of the yacht under consideration, i.e. the heavier model experiences much larger surge forces in following waves than the lighter models. This is an important result because it implies that heavy yachts experience proportionally larger surge forces which must enable them also to surf properly. The relation between the maximum surge force amplitude and of displacement came out to be approximately linear. Also, there is a strong relation between the wave height and the amplitude of the surge force. This relation turned out to be somewhat nonlinear with respect to the wave height. When considering the surge force as a function of the position of the ship with respect to the wave it turned out that the surge force has a high positive peak when the stern is in the vicinity of the wave crest and when the stern is in any other position a lower negative amplitude. This is particulary true for the lighter and the beamier models. Due to the low frequency of encounter and the large nonlinearity an assesment using the period averaged value as is customary in head seas added resistance calculations is not justifiable.

Whether a certain yacht in a particular wave condition will surf or not may only be assessed by lengthy time domain simulations, which due to the strong non-linearities in the forces involved will have a rather 'chaotic' character, i.e. a very small change in the initial conditions may lead to a completely different outcome. Simulations in the towing tank with free moving models (but constant 'thrust' and no sails!) showed however that both the light and the heavy model have good surfing capabilities, quite opposite to 'common belief'. This was also found to be the case in tests on yacht behaviour in breaking waves.[7]

No matter how spectacular and beneficial with respect to speed surfing may seem, it may lead also to very serious problems with directional control of the yacht. This in its turn may lead to broaching of the yacht.

6.6 THE STEERING PERFORMANCE OF THE YACHT

The performance of a sailing yacht is not only characterised by its speed and behaviour in still water and in waves. A very important aspect of its performance is the way it steers. In particular with older designs there have been numerous reported problems with the manoeuvrability and the course keeping qualities of yachts. More recently however, the advent of the modern, fast racers and cruiser–racers with their high powered rigs has also brought problems with controllability, in particular in running conditions. A racing boat that has a high directional stability permits the crew to drive it harder than one that must reduce sail to maintain control. Also for the cruising crew, good steering performance is of interest because such a boat is less exhausting for the crew.

Sailors and designers are well aware of the importance of proper steering qualities of their boats because a boat that oscillates even slightly to either side around a straight line course sails a longer distance and each rudder action and yaw motion creates additional resistance components. This yawing however may be introduced by either the bad steering qualities of the yacht or of the helmsman or both.

The first way to meet the controllability problems is to give the crew more control by devices that can counteract the larger sail forces, i.e. increase the steering power of the ship. This may be achieved by increasing the rudder area and effectiveness or by increasing the distance from the centre of gravity and so increasing its 'leverage'. Another way of improving the steering performance of a yacht may be found in designing the yacht and its controls in such a way that the helmsman as controller of the system spends less time and energy on things that do not contribute to the optimal steering of the ship. This is called the steering compliance of the ship and its controls.

Steering power and steering compliance together form the steering performance of the ship.

Unfortunately calculation of all the forces and moments acting on the hull of a sailing yacht needed for an assessment of the steering properties are difficult to make and experiments to measure them in the towing tank are also very complicated and time consuming and is therefore financially out of reach of most of the design projects.

In the theory described as 'system analysis' the system 'yacht' may be considered to be either an 'open loop' system or a 'close looped' system. When a boat is not steered, i.e. the rudder is fixed, the yachts course (output) is the reaction to forces and moments caused by the fixed rudder input and to external disturbances caused by wind and waves. Such a boat with fixed controls is considered as an open loop system. If a helmsman is asked to steer a ship he will compare the actual course of the ship with the desired course and will control the helm accordingly to counteract the course deviation. All kind of additional information can be used for his reaction, i.e. rate of turn in yaw, inclination, helm angle, sail setting etc. Because of the feedback of the actual course to the helmsman the whole system of yacht and helmsman is called a close looped system.

Most of the research carried out on the steering properties of yachts has been done with the system regarded as an open looped system with fixed controls.

A schematic representation of the open and close looped system of the yacht is depicted in Figure 6.14.

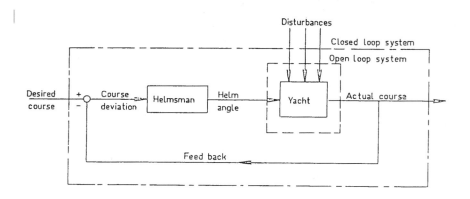

Figure 6.14 *Schematic representation of the open and closed loop system of the yacht.*

A yacht with fixed controls will maintain a straight course as long as the external conditions remain constant. If however the yacht is 'hit' by some kind of disturbance in the external conditions (a wave, a wind gust etc) it will respond in one of two possible ways. The yacht may settle on a new straight course after the disturbance. The yacht is then called fixed control stable. If it does not settle on a new course it may ultimately sail around in a circle and is considered to be fixed control unstable. These two possibilities are depicted in Figure 6.15.

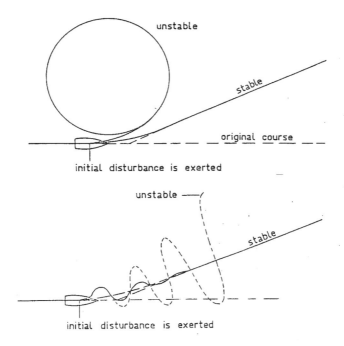

Figure 6.15 *Stable and unstable responses of a yacht to an initial disturbance of its heading.*

Whether a ship will be stable or unstable can in normal ship hydromechanics practise be solved mathematically by solving a set of linear equations describing the motions of the yacht in yaw and sway.

$$(m - Y_{\dot{v}})\dot{\beta} + Y_{\beta}\beta + Y_{\dot{r}}\ddot{\psi} + (Y_r - m)\dot{\psi} = Y_{\delta}\delta_r \qquad [6.6]$$

$$N_{\dot{v}}\dot{\beta} - N_{\beta}\beta + (I_{zz} - N_{\dot{r}})\ddot{\psi} - N_r\dot{\psi} = N_{\delta}\delta_r \qquad [6.7]$$

    v = sway velocity
    ψ = yaw
    m = mass
    Y_v = derivatives of the sway force
    N_v = derivatives of the yaw moment

The solution of the equations [6.6] and [6.7] yields the stability roots which determine the behaviour of a ship after an initial disturbance from the equilibrium condition while the rudder remains fixed. If all roots are real and negative the ship will after the initial disturbance come to a straight path again. If the roots are complex the fixed control behaviour of the ship is oscillatory. The oscillations are

104 *Sailing Yacht Design: Theory*

damped and the ship with fixed control is stable if the real parts of the complex roots are negative. In the case of positive real parts the oscillation is undamped, which means a fixed control unstable ship.

For a sailing yacht this model is not completely correct because the large distance between the centre of effort of the sails and the centre of gravity of the yacht makes a coupled set of equations in sway, yaw and roll necessary. This will not be discussed within the scope of this chapter in great detail but it is an important aspect in describing the directional stability of a sailing yacht.

As shown in Figure 2.1 of Chapter 2, the balance of a sailing yacht is determined by the relative position of the total heeling force on the sails and the sideforce on the hull with appendages when considering the longitudinal vertical plane through the yachts centerline. However when considering the horizontal plane through the yachts center of gravity the moment formed by the driving force component of the sailforces and the 'centre of effort' of the resistance force on the yacht underwater part also has a strong influence on the balance of the yacht. They create a significant yaw moment on the yacht, as shown in Figure 2.3 of Chapter 2.

In particular the later moment is strongly dependent on the heel angle of the yacht because the arm of this moment increases with heel. Both moments are dependent on the forward speed of the yacht because the wave formation in particular at the bow plays an important role in the sideforce distribution and the centre of effort of the hydrodynamic sideforce on the hull.

Without going into much detail here it may be derived for surface ships that the simpliest assessment of whether a straight line dynamic stability is achieved or not is given by:

$$N_\beta \cdot (Y_r - m) - Y_\beta \cdot Nr > 0$$

Y_β = the lift on the hull and appendages due to leeway angle

This force is quite large for a sailing yacht due to the appendages especially designed to generate large lift forces. Effective high aspect ratio widely separated foils contribute the most.

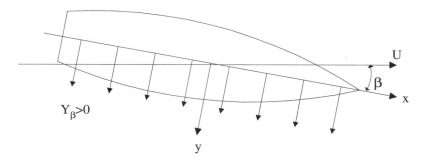

Figure 6.16 Y_β *Lift force on hull and appendages.*

N_β is the moment due to the lift force distribution along the length of the ship when sailed with a leeway angle. If the fore part of the ship dominates $N_\beta > 0$. The shape and position of the appendages largely dominate the sign of this moment, but the dominant contribution arises from the aft placed rudder. The further aft the higher N_r.

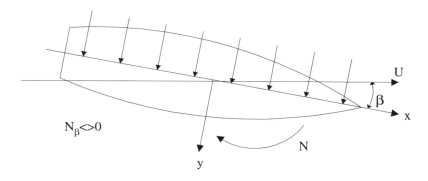

Figure 6.17 N_β *moment due to leeway.*

Y_r is the resultant sideforce due to the velocity of the yaw rotation of the ship. Depending on which part dominates the sign of Y_r may be positive or negative. Here again the aft placed rudder will contribute more than the central placed keel. So the force will most likely be positive. The magnitude of this resultant force, however is generally rather small.

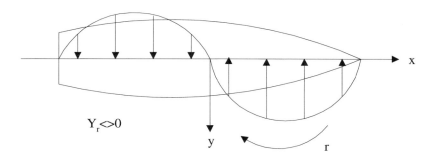

Figure 6.18 Y_r *shift due to rotation of hull and appendages.*

N_r is the moment due to velocity of rotation in yaw in which the forces on fore and aft body both give the same contribution. This is a rather significant moment.

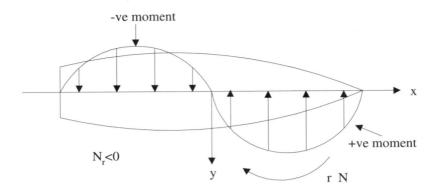

Figure 6.19 *Moment due to rotation of hull and appendages.*

After some mathematical manipulation it can be derived that stability is achieved when

$$\frac{N_\beta}{Y_\beta} < \frac{-Nrr * r}{(mU^2 / Rc) - (Y_r * r)} \quad [6.8]$$

See Figure 6.20 which illustrates that the longitudinal position of Centre of Effort of the sideforce due to leeway should be further aft than the longitudinal position of the Centre of Effort of the sideforce due to the rotation of the ship. The further aft the more stable, although it should be emphasised that no criteria on the amount of positive straight line stability of sailing yachts exists.

Dynamic Behaviour of Sailing Yachts in Waves 107

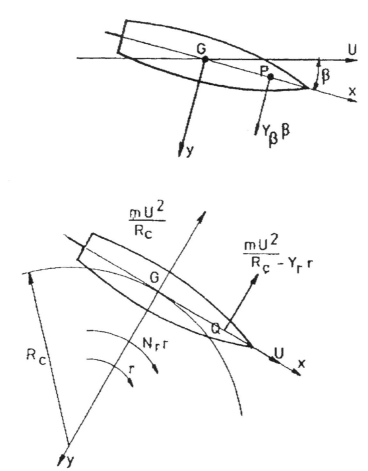

Figure 6.20 *Criteria for fixed controls stability.*

It should be realised that this stability assessment only refers to an initial (small) disturbance from the equilibrium situation. If large disturbances and large resulting deviations of the original stable situation occur important other effects come into play, such as the possible stall of the appendages (keel and in particular rudder) and strong nonlinearities in the forces and moments on the hull. Then the earlier mentioned 'steering power' becomes important also and this may not always be in favour of high aspect ratio foils for keel and rudder, which, as we have seen before, are more vulnerable to stalling effects.

REFERENCES

1. Gerritsma, J. and Beukelman, W. Analysis of the resistance increase in waves of a fast cargo ship, International Shipbuilding Progress, Vol. 19, Nr. 217, 1972.
2. Levadou, M.M.D., Added Resistance in waves of sailing yachts. Report 1032–S Shiphydomechanics Laboratory Delft University of Technology, July 1995.
3. Gerritsma, J. and Moeyes, G. The seakeeping performance and steering properties of sailing yachts, 3rd HISWA Symposium, Amsterdam, 1973.
4. Gerritsma, J. and Keuning, J.A. Performance of light- and heavy displacement sailing yachts in waves, The Second Tampa Bay Sailing Yacht Symposium, St. Petersburg, Florida, 1988.
5. Gerritsma, J., J.A. Keuning and Versluis, A. Sailing yacht performance in calm water and waves, 11th Chesapeake Sailing Yacht Symposium, SNAME, 1993.
6. Keuning, J.A., Terwisga P.F. van and Adegeest, L.J.M. Experimental and Numerical Investigation into Wave Exciting Surge Forces in Large Following Seas. FAST Conference 93, Yokohama, Japan Dec. 1993.
7. Claughton, A and Handley, P., An investigation into the stability of sailing yachts in large breaking waves, University of Southampton, Department of Ship Science Report, No.15, 1984.

CHAPTER 7
VELOCITY PREDICTION PROGRAMS

A. Claughton
Wolfson Unit for Marine Technology and Industrial Aerodynamics,
University of Southampton

7.1 BACKGROUND

Late in the nineteenth century sailing yacht designers were beginning to use the towing tank as an aid. Initially the test techniques were drawn from experience with steam ships and were thus restricted to upright resistance tests. By 1930 it was clear that these simple test methods were inadequate and Professor Kenneth S.M. Davidson, the director of the Towing Tank at the Stevens Institute of Technology in New York, began to investigate a test technique that assessed a yachts performance when heeled and yawed. His paper, 'Some experimental studies of the sailing yacht'[1] marked the start of modern sailing yacht analysis. The paper described a method by which sailing yacht performance could be predicted from separate derivations of hydrodynamic and aerodynamic characteristics. These methods of predicting speed, heel angle and leeway are still the foundations of model testing and performance prediction methods used today.

H. Irving Pratt ocean race handicapping project

In the early 1970s, under the auspices of Commodore H Irving Pratt, researchers at the Massachusetts Institute of Technology, MIT were funded to produce a methodology that would predict the speed of a sailing yacht, given knowledge of its hull, rig and sailplan geometry. The H Irving Pratt project[2] produced the first Velocity Prediction Program (VPP) and covered a huge amount of new ground, including:

- A hydrodynamic force model based on towing tank tests
- An aerodynamic model for the sail forces at all apparent wind angles
- Development of an iterative optimisation scheme which produced equilibrium sailing condition solutions
- Derivation of new methods of race scoring to utilise the seconds/mile handicaps that the VPP predicted for each point of sailing

- Development of a hull offset measuring machine and associated Lines Processing Program (LPP) to pre-process hull shapes prior to introduction into the VPP

IMS

In 1976 the Offshore Committee of U.S. Sailing adopted this computer program as the basis of the Measurement Handicapping System (MHS) to facilitate the equitable handicapping of diverse boat types.

In November 1985 the Offshore Racing Council voted to adopt this system; initially as a second rule alongside the International Offshore Rule (IOR). Finally the International Measurement System (IMS) became the only internationally administered handicap rule. The basic approach of the IMS VPP is described in reference 2, and later improvements to the aerodynamic force model are described in reference 3.

Current VPPs

The IMS VPP, while readily available, is not intended to be, nor is particularly suitable for routine analysis by yacht designers. Its functionality, and the algorithms used for performance prediction are predicated by the needs of a handicapping system, rather than a performance analysis tool. From these MHS and IMS roots several more sophisticated VPP programs have been developed and these offer not only enhanced usability, but also more comprehensive force models, together with the ability to accept force data from experimental and computational results. The need to program a VPP from the ground up no longer exists, except for some very specialised sailing craft, but the very real task facing performance analysts is the creation of more accurate and reliable force models.

The diversity of function and use coupled to the relatively small size of the world-wide market means that there are only a few 'fully functional' commercially available VPP packages. However, a large number of designers and researchers maintain their own 'in house' performance prediction software, which they alone can use with any confidence.

7.2 AIMS OF A VPP

A VPP takes a set of boat parameters and solves for boat speed and associated heel angle for range of true wind speeds and headings, perhaps adjusting the trim or set of sails to achieve the fastest speed. Set against this simple aim the requirements of VPP users are many and varied. For the design of a racing yacht, the VPP is used for a variety of performance issues, for example:

- Systematic series: for sizing and choice of principal form parameters
- Parametric variations: of sail area, rig size, keel draught, stability
- Appendage studies

- Overall rating review – variations of flotation, ballast, sail area, appendage modifications in an effort to improve performance relative to a handicap
- Post processing of towing tank and wind tunnel tests
- Evaluation of the effect of structural changes
- Evaluation of the results from CFD computations

Also the VPP is used for purposes related to design criteria other than competitive performance, for example:

- Appendage sizing through leeway estimates
- Appendage section design through lift coefficient estimates
- Rudder load criteria through estimate of highest speeds
- Jib track leads and angles through leeway calculations
- Resistance estimate for engine sizing
- Reasonableness of rig size and sail area through **Reef** and **Flat** checks

Table 7.1 VPP: Typical user and uses

Amateur yacht designer	VPP used only occasionally at certain stages of producing and developing a design.
Large design house	VPP in constant use for new designs and optimisation.
	VPP may be customised to accept 'In House' data.
Model testing establishment	VPP used as post-processor for Towing Tank and Wind Tunnel Results.
Yacht builders	VPP used extensively for proposal preparation.
	VPP used to optimise hull and rig configurations for new builds.
	VPP used as vehicle for reception of test data.
One off project design groups (e.g. Americas Cup Challenge)	Large design groups have several departments which must be persuaded to talk a common language for performance prediction.
Sailors	Development of target boat speeds and sail use selection charts for inventory selection
Educational Establishments	Provision of a versatile VPP allows students to develop and evaluate new force models, without re-inventing the 'solution' wheel.
	VPP used as teaching aid.
Others (e.g. Sail Makers, Instrument Manufacturers, Race handicappers)	VPPs can provide support to the activities of yacht-related companies.

In addition to the different types of study the users of VPPs have diverse requirements, as shown in Table 7.1. The development of VPPs for different applications are described in references 4–10.

The common thread through these requirements and uses is the need to provide an engineering tool for design work and decision making. A distinction must be made between this stated objective and the emerging theme in computer-aided design to 'automate and optimise'. There is more to design than creating a hull shape and finding out how fast it goes, a meaningful performance assessment can only be made on a fully defined yacht. Consider, for example, evaluating the effect of altering waterline beam as a single variable. When beam is reduced the hydrostatic righting moment is reduced, which then affects the required mast scantlings and hull structural weight. The rig becomes lighter, the centre of gravity goes down, and the righting moment goes up again to some degree. This iterative process that ties form, structures and performance together is critical to proper design development and evaluation.

7.3 METHODOLOGY

All VPPs calculate a steady-state speed V_S and heel angle ϕ solution for the yacht given a true wind speed V_T and heading angle β_T to the wind. The manner of the calculation and the user interface differ widely from program to program. The methodology described in this chapter is based on the 'WinDesign'[*] VPP.

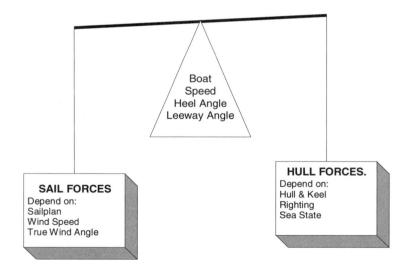

Figure 7.1 *VPP as a force balance.*[28]

---

[*] Developed by Yacht Research International.

Velocity Prediction Programs   113

In general, VPPs have a two-part structure comprised of the **boat model** and the **solution algorithm**. The boat model is often thought of as a black box into which boat speed, heel angle, reef and flat are input. The output is simply two numbers; the difference between aerodynamic driving force and hydrodynamic drag, and the heel angle produced by the heeling moment. It is then the job of the solution algorithm to find a driving force–drag equilibrium, i.e. balance the seesaw in Figure 7.1, and to optimise the sail controls (reef and flat) to produce the maximum speed at each true wind angle.

- Aerodynamic
  For a given wind and boat model variable set (V_T, β_T, V_s, ϕ, *reef, flat*), determine the apparent wind angle and speed that the sails 'see' and predict the aerodynamic lift L and drag D they produce for a given reef and flat. The aerodynamic forces are resolved into thrust, sideforce and heeling moment as described in Chapter 2.
- Hydrodynamic
  Predict the resistance R and righting moment RM the hull produces for the assumed speed and heel angle, given that hydrodynamic sideforce will equal the known aerodynamic sideforce.

It should be noted that currently available VPPs solve only for a balance of force and moment about the X axis. Reasonable extensions of this capability would be to introduce the Mz, yaw moment equilibrium condition, so that sail trimming options or speed and heel values for the yacht that produce excessive yaw moments would be reflected in terms of their influence on speed.

Structure

Figure 7.2 shows that a VPP is not a monolithic structure. The force balance aspect of the process, while computationally intensive is not crucial to the veracity of the predictions. It is the accuracy of the force models and the correct choice of decision and state variables that determines the usefulness of the VPP predictions.

A useful VPP is a system in which the yacht designer can develop and audit the best estimates of the factors influencing the aerodynamic and hydrodynamic performance of a sailing yacht. Consequently the modern VPP must fulfil two roles: firstly to offer a plausible force model that is appropriately sensitive to the input parameters, and secondly, be capable of accepting bespoke force modelling where a user has access to this from towing tank tests or CFD calculations. The VPP can thus be used to carry the data from several sources within a correctly designed and commonly used format.

114 *Sailing Yacht Design: Theory*

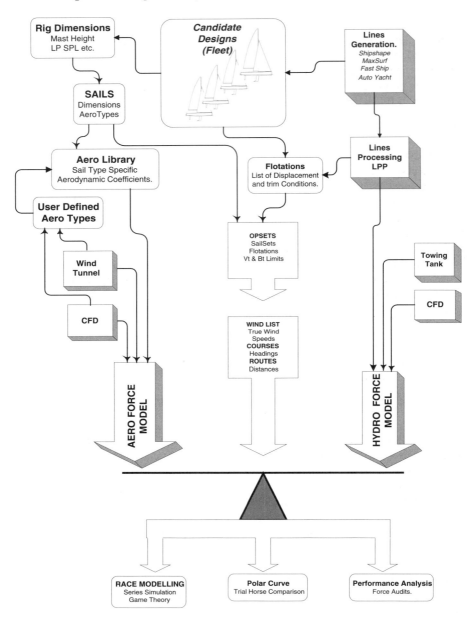

Figure 7.2 *VPP Functional hierarchy.*

The accuracy of the force modelling required within a VPP depends on the use to which the final results will be put. At the preliminary design stage the designers interests are best served by a 'simple' model that requires only a few input parameters. At the other end of the design spectrum a designer might have to evaluate the effect of a detailed change to an appendage which results in a complex change in the interaction between VCG position, viscous and induced drag and possibly a rating change. In this instance the VPP will require a much more detailed force model to produce meaningful results.

In addition to this the VPP must provide for pre-processing of information by other design tools, such as lines fairing packages, and adequate post processing of results in terms of data output, reporting and comparative analyses within the program structure.

Operation and use

There are three sets of parameters required to make a prediction: boat parameters, sailing parameters, and modelling parameters.

Boat parameters

> Physical characteristics of the hull and appendages
> Dimensions of the rig and each sail in the inventory
> Details of the drag-only elements (windage).

Sailing parameters

> Headings, wind speeds, and sea conditions.
> Descriptions of sail combinations for different types of conditions, headings, and wind speeds.
> Flotation conditions to use for different headings or wind speeds.
> Definition of flotation conditions and sails that are used together, termed operational sets

The IMS VPP restricts itself to two operational sets, upwind with mainsail and jib, and downwind with mainsail and spinnaker set, both using a single flotation condition. For yachts that use water ballast, or have multi-masted rigs a wider range of operational sets is required.

Modelling parameters

Underlying the physical description of the boat and sails, there is the computational model. In a multi-functional VPP the user will be required to make decisions about the methods or parameters of the aerodynamic and hydrodynamic modelling for example:

- Will internal force models or experimental data be used?
- Will leeway be included in the calculation of aerodynamic forces?
- Will added resistance in waves be included?
- Will biplane theory be used for keel–rudder interaction?
- What canoe body form factor is appropriate?

The following sections of this chapter describe the approaches adopted for typical force models, solution methods and post processing of results.

7.4 HYDRODYNAMIC FORCE MODEL

The total hydrodynamic drag of the yacht is assumed to be the sum of the following components:

$$R_{TOT} = R_U + R_H + R_I \qquad [7.1]$$

Upright Resistance (R_U) comprising:
    Wave resistance (R_W)
    Appendage viscous drag (R_{Vapp})
    Canoe body viscous drag (R_{Vcb})
Heel drag (drag due to heel alone) (R_H)
Induced drag (R_I)

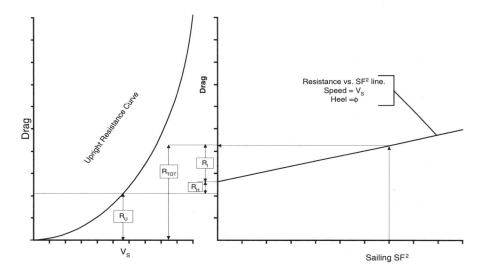

Figure 7.3 *Resistance breakdown for VPP.*

Equation 7.1 shows a convenient way of viewing the calm water resistance of a sailing yacht. The combination of these factors is shown in Figure 7.3. At a speed V_S and heel angle ϕ the resistance value can be determined by the intersection of the resistance against sideforce squared line with the equilibrium sailing sideforce line shown as a dotted vertical line in Figure 7.3. The requirement of the hydrodynamic force model is to determine these three resistance components. This is typically done using a series of formulations, or results from physical testing or computer modelling.

Viscous drag

Canoe body

The canoe body friction and form drag is usually assumed to be proportional to static wetted surface area at each heel angle, using a C_F calculated from an appropriate friction line and reference wetted length. Conventionally a single value of form factor is assumed regardless of heel angle, but a more complete approach,[11] is to adopt a F_N and heel angle dependence for wetted area, length and form factor.

Appendages

The appendage viscous drag $R_{v,a}$ may be calculated using the following expression:

$$R_{v,a} = q(1+k_a) C_{f,a} S_a + c_q L_a^2 \qquad [7.2]$$

The frictional resistance coefficient for the keel fin, bulb and rudder may simply be based on the ITTC formulation and the form factor derived from published data (e.g. Hoerner[12]). More commonly, CFD codes are used to determine the viscous resistance of the keel fin and rudder. The dynamic pressure q is reduced for the rudder since it is partially in the wake of the canoe body and keel. The second term in the drag expression accounts for the increase in form drag when the appendage is generating lift.

Residuary resistance

Expressions for determining residuary resistance need not be very complex, indeed the first MHS VPP formulation[2] (equation [7.3]) had only three coefficients (a_1–a_3) defined at eight values of Fn. This is possible because the major driver of residuary resistance is displacement which occurs in the denominator of the residuary resistance coefficient.

$$\frac{R_r}{\Delta} = a_1 + \left(\frac{B}{Tc}\right)^{a_2} \frac{Cv}{\sqrt{Cv^2 + a_3}} \qquad [7.3]$$

Cv = volume/L^3
B = beam
Tc = canoe body draft
R_r = residuary resistance
Δ = displacement

Contributing to the simplicity of this early formulation was the fact that the characteristic length used in the derivation of Fn was determined from the second moment of area (LSM) of the sectional area curve of the immersed part the hull, as shown in equation [7.4].

This approach is still used today in the IMS to avoid the distorting effects of point measurements. The regression equation [7.3] tackled only the effect of length/volume ratio and beam/draft ratio. The effect of nuances of volume

118 *Sailing Yacht Design: Theory*

distribution within the canoe body were handled by the LSM. Over time residuary resistance equations have been developed to include other form parameters such as Cp and LCB as described in Chapter 5, and this has made the characteristic length inferred from volume distribution less appropriate.

$$LSM = 3.932 \sqrt{\left(\frac{\int x^2 \sqrt[4]{s}\,dx}{\int \sqrt[4]{s}\,dx}\right) - \left(\frac{\int x\sqrt[4]{s}\,dx}{\int \sqrt[4]{s}\,dx}\right)^2}$$ [7.4]

    s = element of sectional area attenuated for depth
    x = length in the fore and aft direction

The method of determining a characteristic length value for use in the calculation of Fn depends on the use to which the result will be put. For the analyst the use of the static waterline length is a sensible choice, and the Delft systematic series formulae are based on this. However the characterisation of length as part of a measurement rule requires this process to carry two further burdens, firstly the creation of a length that is not fixed to a single point, and hence susceptible to exploitation by designers, and secondly to take account of the effect of overhangs on the hull resistance as discussed in reference 14.

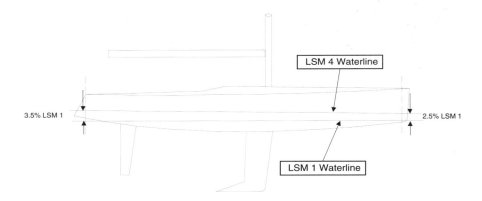

Figure 7.4 *Determination of IMS L from LSM values at 3 flotation waterlines.*

As speed increases the yacht sinks down and the heights of the bow and quarter wave increase, immersing the overhangs. The IMS VPP method of accounting for this effect is by calculating 'L' as a mean of three LSM's at different waterlines, shown in Figure 7.4.

    L = 0.3194 × (LSM1 + LSM2 + LSM4) [7.5]

        L = IMS length
    LSM1 = second moment length for the yacht in Sailing Trim floating upright
    LSM2 = second moment length for the yacht in Sailing Trim floating with

LSM4 = 2° heel
second moment length for the yacht in a deep condition sunk 0.025*LSM1 forward and 0.0375*LSM1 aft, floating upright

An alternative approach described in reference 5 is to consider a high-speed length L_{hs} which is the physical wetted length in a deeper or sunken flotation condition. At a given 'high speed' Froude number Fn_{hs} we assume the bow and stern waves have risen to the specified deeper waterline. Between $Fn = 0$ and $Fn = Fn_{hs}$, length is increased smoothly from Lwl to L_{hs} in a prescribed way.

Appendages

References 14 and 15 describe how the appendages contribute to the residuary resistance component, for example adding a rudder at the aft end of the waterline, or a bulb keel modify the residuary resistance. Chapter 5 describes these effects in more detail, but they are difficult to implement in the general case of a VPP. The IMS VPP contains a function that accounts for the contribution of keel volume, and the determination of length from the integration of sectional area is sensitive to rudder volume.

Drag due to heel

As a yacht heels the immersed hull becomes asymmetric and modifies the wave resistance, viscous drag and generates sideforce. Some of these effects are included elsewhere, specifically:

- Variation in canoe body wetted surface with heel is accounted for in the canoe body drag calculations, although fluctuations in form factor are often carried within the residuary resistance component
- Effects of heel on induced drag are accounted for in the induced drag calculations

The remaining effects are classified here as heel drag and considered as three components, the first two of which are considered wave-making effects and based on the following:

- Change in wetted length from upright as the boat heels
- Asymmetry or change of effective camber of the hull

The third component is a viscous effect and is due to a variation in canoe body form drag. Typical heel drag formulations are given in references 4, 8 and 16, and the lower part of Figure 7.5 shows a typical variation of heel drag with heel angle and F_N.

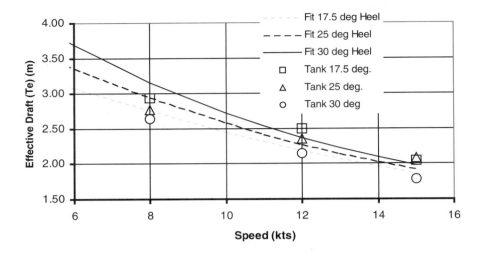

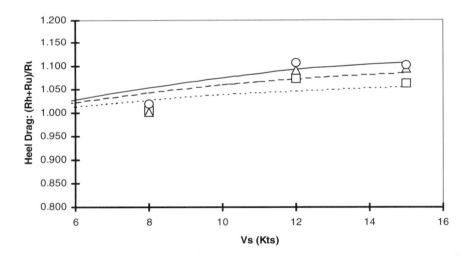

Figure 7.5 *Typical variation of drag due to heel and effective draft with speed and heel angle.*

Sideforce and induced drag

Chapter 5 describes how induced drag arises and how it can be expressed in terms of an effective draft, which approximates to the yachts physical draft. Conventionally, VPPs use the following derivation of induced drag (D_I) based on an effective draft (Te).

$$D_I = \frac{Heeling\ Force^2}{Te^2 \pi \rho V^2} \qquad [7.6]$$

(Note this equation differs from equation [5.4] in Chapter 5 due to the assumed Aspect Ratio being based on a hull and keel that are reflected about the waterplane.)

If the yacht behaved as an aeroplane then, for a given wing geometry, the Te would remain constant regardless of speed and heel angle. However, because the yacht operates at the air–water interface the pressures on the keel root and rudder create waves which add to the drag. This extra drag is indistinguishable from that due to the shed vorticity and can be construed as giving the induced drag a sensitivity to speed and heel angle. The upper part of Figure 7.5 shows a typical variation of Te with speed and heel angle, but the VPP is hard pressed to model these complex interactions of sideforce and wave-making except in very general terms. As a first step the effective draft is usually calculated from the draft of the yacht, modified to account for the wake contraction caused by the presence of the canoe body.[2]

$$Te^2 = T_{max}^2 - \frac{2Ax}{\pi} \qquad [7.7]$$

T_{max} = maximum draft
Ax = hull cross-sectional area amidships

Modern yachts increasingly configure the rudder to carry a large proportion of the hydrodynamic sideforce. Consequently it is now common to include the effects of the rudder by treating the induced drag from the keel and rudder combination using biplane theory.

Effect of keel geometry

Whilst the previous section has shown that effective draft is strongly dependant on the vessel's maximum draft, the shape of the keel tip also has an influence. On traditional yachts this discussion would centre around the use of V or rounded tips on keels which were similar to simple wings.[17] However one of the strongest trends in modern yacht design is to reduce the lateral area of the keel fin, and to carry the ballast in a keel tip bulb. The presence of such a large, essentially non-lift producing body on the bottom of the keel reduces the keel's effective draft (Te). Figure 7.6, based on data from reference 18, shows the extent of this Te reduction. Keel wings and detailed modification of the bulb geometry by introducing mid section chines and horizontal trailing edges, can offset the damaging influence of the bulb on the Te.

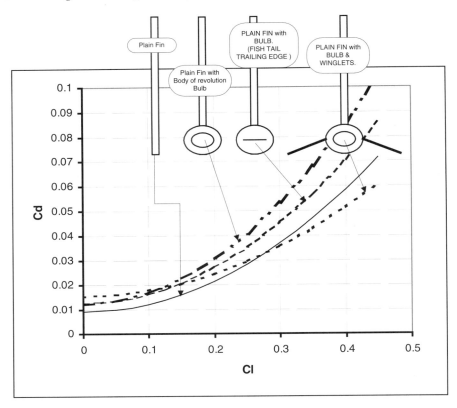

Figure 7.6 *Cd versus Cl^2 for different keel tip geometries.*

Figure 7.6 shows how adding wings to the tip of the keel allows some of the energy from the tip vortex to be recovered, thereby reducing induced drag. However the wings themselves add to the viscous drag of the appendages. As described in Chapter 2 the sailing sideforce depends on the righting moment and sailplan of the yacht. Thus there is no value in adding wings to a yacht whose sailing lift coefficient in normal upwind conditions falls below the cross over Cl value, approximately 0.29 in Figure 7.6.

Naturally the VPP will be used to assess the yachts 'round the course' behaviour, but Figure 7.6 is an important picture to keep in mind when considering the design of additions intended to affect Te.

Speed and heel effects

Figure 7.5 shows the typical behaviour of a conventional sailing yacht in that Rh increases and Te reduces as speed and heel angle increase. This pattern can however be modified by the use of a bow rudder. Table 7.1 compares a hull fitted with a conventional keel and rudder with the same hull with a bow rudder and deeper aft rudder added. At low speed this has no effect on Te, but at higher speeds the pressure distribution on the forward rudder interacts favourably with the hull generated bow wave and resistance is reduced, giving an increase in Te. When devising expressions for Te and Rh these effects are hard to predict without some form of test data.

Table 7.1 Effect of full depth rudder and bow rudder on effective draft and heel drag

Speed (kts)	Conventional Te (m)	Modified Te (m)
10	2.62	2.62
12	2.37	4.48
14	2.10	5.76

Leeway

Leeway is not an independent variable and does not directly enter into the solution. The hydrodynamic sideforce balances the aerodynamic sideforce and the yacht will assume some leeway angle to produce such a sideforce. However leeway values are used to compute the component of viscous drag on the appendages when they are lifting, and used to evaluate the gap between the keel and the rudder when biplane theory is used to calculate effective span. Leeway calculations generally include the effects of hull size, beam, canoe body draught and the effective spans of keel and rudder. It is Froude Number and heel angle dependent in the same way as effective draft.

Righting moment

The LPP uses the calculated VCG and volumes of canoe body and appendages to calculate GZ curves for each of the specified flotation conditions.

The effect of forward speed and the associated pressure fields around the hull modify the GZ from the statically calculated value. Traditionally it has been assumed that GZ is reduced because of the effect of the mid ship wave trough. However on more modern hull shapes with fuller ends and beamier aft overhangs at higher speeds the bow and stern wave crests may produce a GZ increase. Thus care must be exercised in using calculations of dynamic righting moment that are extrapolated from low speed tests. The wave patterns and modified hull surface pressures arising from hydrodynamic lift also modify the VCLR position.

VCLR and CLR

The LPP traditionally calculates the hydrodynamic vertical CLR using a simple expression, for example:

$$Zcp = 0.42\ Tmax - k\ Tc \qquad [7.8]$$

Zcp = vertical centre of pressure below waterline
Tmax = maximum draft
Tc = canoe body draft
k = constant derived from hull parameters

The calculation of CLR is typically done outside the VPP as a check that steady state rudder angles are not too high or low at the equilibrium conditions. Reference 19 describes a method of calculating actual CLRs as opposed to centroids of the lateral area associated with traditional lead and lag calculations.

Added resistance in waves

Added resistance in waves (R_{AW}) is usually handled in one of two ways.

1. By introducing a computationally[20] or experimentally[21] derived R_{AW} response amplitude operator (RAO) into the VPP and combining this with a sea spectrum to yield an added resistance value. It is usual to include within the VPP a characterisation of the wave spectrum significant height and modal period, expressed as a function of true wind speed (V_T) so that an appropriate added resistance can be calculated at each wind speed.
2. Implementation of a parametrically based calculation using the physical characteristics of the yacht.

The IMS VPP contains a parametric model based on an amalgam of model test results and CFD calculations. The R_{AW} is calculated from an equation which is a function of the hull parameters and is intended to yield a value that is appropriate to an average wave climate for each V_T value. Naturally this is not correct all the time as the wave spectrum is sensitive to wave fetch and wind duration. Nevertheless it is a fundamental part of the IMS as it is the vehicle by which the effect of hull inertia variations are handicapped. It is usual within a VPP to be able to isolate the R_{AW} effects, but in the IMS VPP, because it is a handicap system, the R_{AW} is always activated. Chapter 6 describes the regression equations to calculate R_{AW} from the basic hull parameters derived from the Delft Systematic Yacht Hull Series.

Tank testing

There are two distinct types of model testing carried out in support of VPP calculations.

1. Standard Series tests to derive parametric regression data for series, such as the Delft Series described in Chapter 6
2. The comparison of different designs as candidates to be built for a particular competition

Neither of these approaches has proved entirely satisfactory to yacht designers seeking to improve their designs. There is sometimes criticism that the models tested as part of a true standard series are not refined hull shapes, due to the constraint of having to achieve particular parametric values. Conversely the analysis of tests on alternative designs, which were drawn without reference to some thematic variation of parameters, is difficult because it is hard to isolate the individual effects. Between these two extremes a new approach of testing 'candidate designs' has begun to find favour. While the test models might represent steps in a parametric variation, each model could, if required, be built and sailed with some prospect of success. This means that the designer must be happy with the hull form, and the choice of keel and sailplan must be appropriate to achieve the required rating.

Analysis and fitting methods

There are conflicting requirements in making the step from experimentally derived data to a mathematical model which the VPP can evaluate. Conventionally, experimental results are determined at discrete conditions (i.e. fixed values of boat speed, and heel angle as described in Chapter 14), whilst the VPP requires a continuous definition of behaviour. To make this step some form of regression analysis must be performed to derive coefficients in an equation that describes the yacht's behaviour. The goodness of fit of the results from the regression equation to the measured data can always be improved by increasing the number of terms or coefficients. This, however, creates two problems; firstly, as the number of terms increases it becomes harder to ascribe a physical significance to each one, and, secondly, it becomes impossible to extrapolate the coefficients reliably beyond the tested matrix.

The total resistance breakdown based on towing tank results was shown in Figure 7.3. Having derived this data by testing and analysis, described in Chapter 14, the type of post processing for transmission to a VPP will depend on the aim of the tests.

Either this will be an evaluation of a candidate design, in which case the aim is to produce an equation for total resistance that fits the test data as well as possible and offers plausible force values away from the tested points. It is sometimes useful to introduce only the residuary resistance component into the VPP because in this way geometrically similar variations of the tested hull, or small variations in appendages can be quickly evaluated in the VPP using calculated viscous components of resistance.

Or the data are to be used for creating a model of a particular component of resistance, in which case the data must be decomposed into the viscous and residuary components prior to undergoing statistical analysis with the other members of the series.

Fitting methods
When evaluating candidate designs and their close variations, either a scheme that interpolates between test points (e.g. reference 16) or a 'functional fit' method, as discussed in reference 22 (and exemplified by references 5 and 8), may be used to derive an expression for total resistance that can be evaluated by the VPP. The

functional fit method usually offers the most satisfactory results for the following reasons:

- Functional fits are able to fair through the scatter of data points and if correctly configured do not mask hydrodynamic nuances in the test data set, particularly if the test points are weighted according to their closeness to the sailing sideforce.
- The function does not need to be very complicated, since the upright resistance curve carries the major burden of defining the resistance characteristics. The function therefore need only reflect the speed and heel angle dependence of the drag due to heel and the induced drag. Figure 7.5 shows the fit of a simple equation, of the form shown below, to the experimentally determined values which are derived for discrete combinations of speed and heel angle.

$$R_{TOT} = R_U + R_U(c_1 + c_2 V_S + c_3 V_S^2) \phi^{c_4} + FH^2/V_S^2 (c_5 + c_6 V_S + c_7 V_S^2 + c_8 \phi^2) \quad [7.9]$$

$c_1 - c_8$ = coefficients determined by regression analysis

By using a simple expression the coefficients may be constrained to ensure consistent behaviour of the predicted resistance outside the range of the test points, and they may be manipulated to engender familial relationships between members of a test series. Judicious human touch can ensure a rational and plausible extrapolation scheme from sparse data sets.

Figure 7.5 shows test points (combinations of speed and heel angle) that mirror as far as possible the actual speeds and heel angles of the yacht when sailing. Nevertheless trends of R_H and Te can still be discerned throughout the full speed and heel angle range because the curves of Te and R_H have been forced to have plausible behaviour at the high and low speeds that lie outside the test matrix. The test matrices required to ensure a good blind statistical fit to more complex regression analyses with many terms require the sailing conditions to be bracketed to allow reliable interpolation, hence a lot of testing is expended on non-sailing conditions.

All these considerations apply to fits of either total or residuary resistance for the consideration of candidate designs.

Detail discussion of analyses to isolate parametric effects is too specialised a subject to be dealt with adequately in this chapter. Reference 4 offers comprehensive descriptions of various studies. If analysis is to be extended to consideration of residuary resistance modelling of a more general nature then some further caveats apply:

- Isolating residuary resistance is much more difficult than deriving an accurate full scale prediction of resistance from a single model test, due to the imponderables associated with the viscous resistance component. The residuary resistance can fluctuate by several percent or more at low speeds, depending on the friction line, wetted length and form factor used in the scaling process.
- Once a generic formulation for residuary resistance has been derived, features of the tests, such as the sail trimming moment, become much more arbitrary, as the

rig height for a design may vary from that used to determine the sail force trimming moment used in the tests.
- The method of calculating reference length values will influence Froude No.

Wind tunnel tests

Wind tunnel tests are usually used to determine the relative induced and viscous drag characteristics of keels and their ballast bulbs, which may or may not be fitted with wings. The results of this type of test are most commonly expressed in the VPP simply as an effective draft and associated Cd0 value. Unless the tests are conducted at full scale Rn then the viscous component of the resistance must be scaled from model to full scale, just as is done with towing tank results.

CFD

CFD may deliver either all or part of the hydrodynamic force model. It is normally used to provide detailed analysis of particular aspects of the design, for example appendage viscous resistance from boundary layer calculations,[23] or keel effective span calculations from a panel code.[24] By correctly configuring the user interface of the VPP the internal algorithms may be used to complete the hydrodynamic force model.

7.5 AERODYNAMIC FORCE MODEL

Modelling a set of sails flying together is relatively simple as forces may be reduced to coefficient form based only a reference area and the dynamic head. Set against this apparent simplicity is the fact that different sails are set for each point of sailing, and sails may be trimmed differently to adjust their aerodynamic characteristics as wind strength and point of sailing change. Once again for ease and flexibility of calculation and as an aid to understanding the physics of sailing, it is possible to reduce the total aerodynamic force to a series of individual components.

- Lift
- Viscous drag due to air friction over the sails
- Viscous drag associated with alteration of the boundary layer when lift is produced
- Drag due to separation around bluff bodies such as the hull mast and rigging wires, usually termed 'windage' drag
- Induced drag due to the generation of lift
- The position of vertical centre of effort

128 *Sailing Yacht Design: Theory*

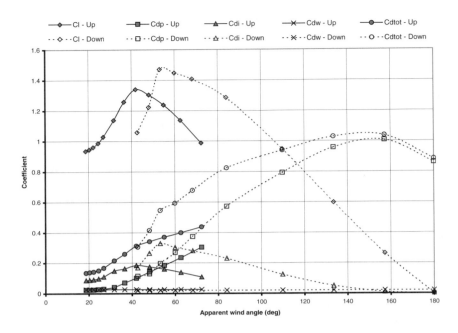

Figure 7.7 *Total aerodynamic forces with components.*

The relative magnitude of the components and their variation with apparent wind angle (β_{AW}) is shown in Figure 7.7.

Sails

The fundamental components of an analytical aerodynamic model are the individual sails. In a typical aerodynamic model a single sail is characterised by the following parameters shown in Figure 7.8:

- Sail area
- Centre of effort height above the sails datum
- Vertical span
- Cl_X and Cd_P versus β_{AW} envelope. (Maximum lift coefficient & Parasitic (viscous) Drag coefficient versus Apparent Wind Angle.)

These values vary with the type of sail, for example mainsail, jib, and spinnaker. The VPP also requires knowledge of how to treat these sails when set together so that calculations of blanketing effects and calculations of combined effective span are correctly carried out.

Velocity Prediction Programs 129

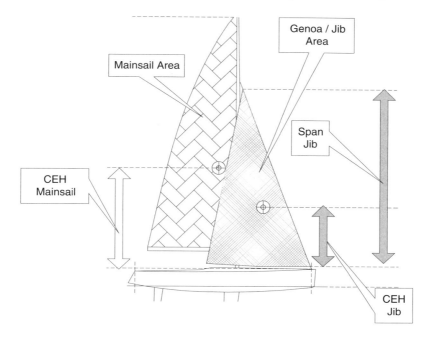

Figure 7.8 *Sail and sailplan geometry definition.*

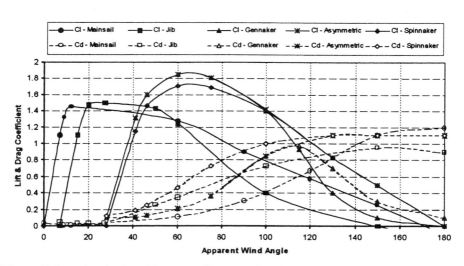

Figure 7.9 *Individual sail force coefficients taken from IMS VPP.*

Sail force coefficients

Figure 7.9 shows the individual two dimensional coefficients for the 5 sail types supported by the IMS.[25] The characteristics of the mainsail and jib and spinnaker were derived empirically when the sail force model described by Hazen[3] was introduced. The coefficient values show typical effects:

- As β_{AW} increases, a rapid rise in lift to a peak value prior to the onset of separation and stall.
- The sails 'fill' at different apparent wind angles, reflecting the different sheeting arrangements and shapes of the sails.
- At an apparent wind angle of 180°, approximating to an angle of attack of 90°, the lift has declined to zero and the drag coefficient increased to 1.0

Sailsets

A combination of sails that are set together may be referred to as a Sailset. The force model for the Sailset can be developed from an aggregate of its component sails, or it can be defined with actual data from experiments or some other source.

Lift and drag and CE height

Aggregate maximum lift and linear parasite drag coefficients are the sum of each sail component's contribution normalised by the reference area:

$$Cl_x = \Sigma\, Cl_{xi}\, B_i\, A_i\, /\, A_{ref} \qquad [7.10]$$

$$Cd_p = \Sigma\, Cd_{pi}\, B_i\, A_i\, /\, A_{ref} \qquad [7.11]$$

The subscript i refers to individual sails and the function B_i is a 'blanketing' function that may be used to introduce sail interaction effects. It is usually equal to 1.0, meaning no blanketing effect.

$$A_{ref} = \Sigma\, A_i \qquad [7.12]$$

The centre of effort height Z_e is evaluated by weighting each sail's individual centre of effort height by its area and partial force coefficient (comprised of lift and linear component of parasite drag):

$$Z_e = \Sigma\, z_{ei}\, (Cl^2_{xi} + Cd^2_{pi})^{1/2}\, B_i\, A_i\, /\, A_{ref}\, /(Cl^2_x + Cd^2_p)^{1/2} \qquad [7.13]$$

The quadratic parasite drag coefficient K_q is the sums of the individual sails contributions:

$$K_q = \Sigma\, k_{qi}\, Cl^2_{xi}\, B_i\, A_i\, /\, A_{ref}\, /Cl_x^2 \qquad [7.14]$$

Effective span and induced drag

The effective span b_e is calculated according to biplane theory for pairs of sails. This applies to the mainsail–headsail combination, or the mizzen–mizzen staysail combination. For a two-masted yacht, the effective spans from the mainmast and mizzenmast sail-pairs are then combined to yield an overall effective span.

The effective span b_e for a single 'independent' sail is given by its physical span times a factor. For a sail pair, such as the genoa and mainsail, the individual effective spans and knowledge of the lift coefficients for each sail is used to derive a combined effective span for the two sails working together. This can be derived by starting with the following equation from Glauert[26] for total induced drag of a biplane:

$$Cd_I = (A_{ref}/\pi) [Cl^2_1 / b^2_{e1} + 2\sigma Cl_1 Cl_2 / (b_{e1} b_{e2}) + Cl^2_2 / b^2_{e2}] \quad [7.15]$$

subscripts 1 and 2 reference the two sails and σ is a coefficient, given by Glauert as a function of gap between the sails and their half spans.

Knowing that the total lift coefficient C_L is the sum of C_{L1} and C_{L2} from Glauert's equation above we can derive an overall effective span b_e that satisfies the induced drag expression for an equivalent single sail:

$$b_e^2 = (A_{ref}/\pi)(Cl^2/Cd_I) \quad [7.16]$$

Total lift and drag

The total sail lift and drag coefficients are given as:

$$Cl = \text{flat reef}^2 Cl_x \quad [7.17]$$

$$\begin{aligned} Cd &= \text{parasite drag + quadratic profile drag + induced drag} \\ &= \text{reef}^2 Cd_p + K_q (\text{reef flat } Cl_x)^2 + (\text{reef flat } Cl_x)^2 A_{ref}/(\pi b_e^2) \end{aligned} \quad [7.18]$$

In which the reef parameter represents a linear reduction in span or chord, but it is squared in these equations. Thus reef = .9 means sail area is reduced to 81% of reference area. The *flat* parameter characterises a reduction in sail camber such that the lift is proportionally reduced from the maximum lift available. Thus flat = 0.9 means 90% of the maximum lift is being used.

The thrust Ct and heeling force Ch coefficients are resolved from the lift and drag coefficients. The heeling moment coefficient Cm is sideforce taken as a moment around the waterplane.

$$Ct = Cl \sin(\beta_A) - Cd \cos(\beta_A) \quad [7.19]$$

$$Ch = Cl \cos(\beta_A) + Cd \sin(\beta_A) \quad [7.20]$$

$$Cm = Ch (Z_e + Z_{ref}) \quad [7.21]$$

132 *Sailing Yacht Design: Theory*

The apparent wind angle β_A is relative to the yacht's track and the heeled mast plane. It is evaluated at the calculated height of the centre of effort, as is the dynamic pressure q.

Heel and leeway in aerodynamic modelling.

Leeway is not used for modifying the wind triangle geometry. When leeway is actually incorporated in the wind triangle, empirical sail coefficients, for example those used in the IMS VPP, are no longer correct and must be modified to remove the built-in mean leeway. Lift and drag coefficients developed through wind tunnel testing, CFD, or analytical means should differ slightly from empirically derived coefficients because of the different assumptions for leeway.

Windage

Windage is the aerodynamic drag from components of the rig, hull, superstructure and crew which produce negligible lift and can therefore be treated in a simplified way. A wing mast or other element which provides lift as well as drag would be better modelled as a special type of sail. A typical windage force model is based on establishing a set of windage components, e.g. hull topsides, crew, rigging, and mast. Each element is ascribed a drag coefficient and frontal area, that is a function of apparent wind angle.

The windage drag of the mast is influenced by the presence of the mainsail attached to its aft edge and therefore alternative drag coefficients may be specified to reflect the higher drag of the bare mast exposed when the mainsail is reefed.

Wind tunnel testing

Results from wind tunnel tests on complete Sailsets (e.g. reference 27) allow the accuracy of this simple characterisation of sail behaviour to be judged. Section 4.6 describes how sails may be set for upwind sailing, both for the maximum drive condition and when eased to reduce heeling force. The plot of this data as Cd vs. Cl^2 shows a linear reduction of Cd as Cl^2 is reduced. This is the behaviour described by equation 7.18. The only area where the idealised model and the physical behaviour differ is in the setting of the sails with some separation drag to achieve the maximum drive condition. This manifests itself as a deviation of the high Cl data points from the straight line, while the idealised force model must retain a linear decline from its Cl_{max}. This is shown in Figure 7.10, which is based on data for a single apparent wind angle taken from Figure 4.10.

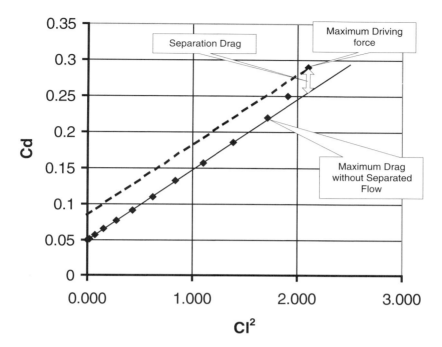

Figure 7.10 Cd versus Cl^2 data from typical wind tunnel tests showing increase of separation drag to produce maximum driving force.

Analysis and fitting methods

Figure 7.11 shows the mechanism by which wind tunnel tests on a complete rig may be introduced into the VPP.

The Cl_{max} and corresponding Cd are described as continuous curves against β_{AW} The reduction of Cd as sail flattening takes place is described by the effective rig height (He).

$$Cd = \frac{A\,Cl^2}{\pi He^2} \qquad [7.22]$$

Because the maximum drive condition may be associated with some separated flow, the actual decline of Cd as Cl_{max} is reduced would follow the dotted line shown in Figure 7.10. This can be dealt with in the VPP by either defining an alternative equation for the Cd vs. Cl^2 relationship or defining two sets of coefficients for the sail set, one for Cl_{max} and a second pair of curves defining the maximum Cl achievable without separated flow, as shown in Figure 7.11. Using this lower curve the flat term will correctly model sail force behaviour as the sheets are eased. In order to utilise this approach of using multiple coefficient sets for a single set of sails it is necessary for the VPP to be able to accommodate multiple sail sets within its operating hierarchy, this is possible for instance using the WinDesign VPP.

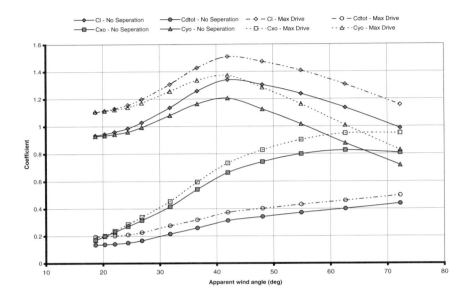

Figure 7.11 *Envelope curves of Cl and Cd versus apparent wind angle.*

When analysing wind tunnel tests the windage from hull and mast and rigging may be included with the derived coefficients, or more usually, the bare hull and mast windage drag is subtracted from the measured results, and the internal VPP windage force model used in its stead. This most easily accounts for any excess drag incurred on the model due to Rn effects, and also allows the sail forces to be applied in a general case, and allows the reef function to more accurately reflect real behaviour.

Sailing dynamometer

The original IMS sail force coefficients were derived by inferring sail forces from the observed performance of a hull whose hydrodynamic forces had been predicted from towing tank tests, based on the simple force and moment balance described in Chapter 2. At the same time attempts were made to determine sail forces from a fully rigged yacht hull tethered by load cells. This approach was unsuccessful due to the wild oscillations of the hull robbed of the damping from the keels forward motion. A halfway house between these two approaches is the 'sailing dynamometer' described by Milgram.[28] This approach involves mounting the whole mast, rigging base and sail sheeting points on a separate sub-frame, and mounting this via load cells onto the yacht hull. In this way all the forces that the aerodynamic elements of the yacht exert on the hull may be measured. It is the sailing equivalent of the power craft shaft torque trial, except that instead of a single torque measurement the researcher is faced with data from several force blocks. This type of measurement models perfectly the way in which sails may be trimmed, the oscillations of wind and sea, the effect of elastic deformation, and the behaviour of offwind sails without having to correct for blockage. The price, however, is high, in that the tests are time consuming to carry out and must be done by technicians who are good sailors. Also the monitoring of sail shapes and a meaningful determination of the wind speed require the presence of support boats. Nevertheless the research by Milgram demonstrates that it can be done if you have the will, and the budget.

7.6 SOLUTION AND OPTIMISATION ROUTINES

Contemporary velocity prediction programs treat the yacht performance prediction problem as one of rigid body equilibrium, where the resultant of all forces and moments acting on the boat are zero. The position of the yacht depends on six degrees of freedom. Of these six, two are routinely considered in a VPP: longitudinal force and rolling moment. Vertical force and pitching moment are assumed to be equilibrated by hydrostatics, and yawing moment is assumed to be equilibrated by rudder moment. Sideforce is equilibrated when the yacht assumes an angle of leeway to develop the hydrodynamic sideforce to balance the aerodynamic sideforce. There are numerous methods for solving the equilibrium equations, those described here are employed in the WinDesign, WinCat and WinGold VPPs written by Clay Oliver.

Ballasted monohull

The two equilibrium equations can be written as functions of five variables:

V_s = boat speed (or $V_{mg} = V_s \cos \beta_T$) [7.23]

ϕ = heel angle
β_T = true wind angle
reef = proportion of sail dimension over maximum sail dimension
flat = sail lift/maximum sail lift

The first two equations we know must be solved are for longitudinal force and roll moment:

ΣFx = thrust − resistance = 0 [7.24]

ΣM_x = heeling moment − righting moment = 0 [7.25]

We have two equations at this point and five unknowns. The three other required conditions are derived from our goal to find values for three of the five variables that will maximise our speed, or speed-made-good. We thus have two *state* variables V_s and ϕ and three *decision* variables β_T, *reef* and *flat*.

This would be the case solving to find the β_T which would optimise V_{mg}. Solving to find the best speed for a given β_T there are two decision variables *reef* and *flat*. Consider this simplified four-variable case. For notational simplicity let F represent ΣF_x and M represent ΣM_x, v for V_s, r for *reef*, h for *heel* and f for *flat*.

Now consider small variations in the four variables dv, dh, dr, df away from some point where our two equilibrium equations F = 0 and M = 0 are satisfied. In order for these equations to remain satisfied, the total differentials of F and M must remain zero:

$$dF = F_v\, dv + F_h\, dh + F_r\, dr + F_f\, df = 0 \qquad [7.26]$$

$$dM = M_v\, dv + M_h\, dh + M_r\, dr + M_f\, df = 0 \qquad [7.27]$$

In which the subscripts represent partial derivatives. By combining these two equations and eliminating d_h, we can write:

$$dv = -[(F_r M_h - F_h M_r)/(F_v M_h - F_h M_v)]dr - [(F_f M_h - F_h M_f)/(F_v M_h - F_h M_v)]df \qquad [7.28]$$

This can be thought of as

dv = [partial derivative of v with respect to reef holding flat constant]dr + [partial derivative of v with respect to flat holding reef constant] df

$$dv = (\partial v/\partial r)dr + (\partial v/\partial f)\, df \qquad [7.29]$$

When v is a maximum with respect to reef, then its partial derivative $\partial v/\partial r$ will be zero. This requirement yields:

$$(F_r M_h - F_h M_r) = 0$$

Which becomes a third equation to satisfy. The fourth equation follows from $\partial v/\partial f = 0$, which means

$$(F_f M_h - F_f M_r) = 0$$

We now have four equations to be solved simultaneously to determine the four unknowns (for the case of fixed β_T). A multidimensional Newton-Raphson iterative solution is used for this purpose, which includes inequality constraints on reef and flat so that they do not go outside the range of 0.0 to 1.0.

Sailing dinghy

In reaching an equilibrium sailing condition for a conventional monohull the VPP solution algorithm is helped considerably by the fact that as heel angle increases so does righting moment, and hence a small increase in heeling force will produce a small increase in heel angle, and essentially heel angle is limited by the decline in sailing speed if heel becomes too great.

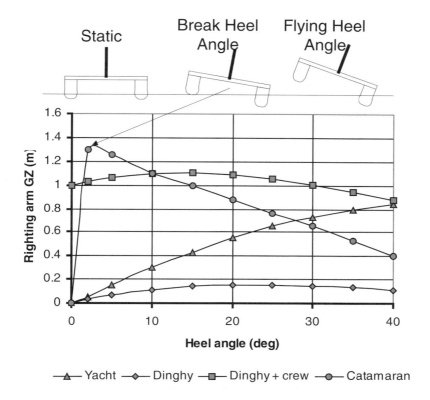

Figure 7.12 *Righting arm (GZ) curves for a yacht dinghy and catamaran.*

In the case of a sailing dinghy with trapezing crew the maximum righting moment occurs at a low heel angle, as shown in Figure 7.12, and this, coupled to the open nature of the boats leads to them being sailed by their crews relatively upright. Consequently in the VPP a limit of sailing heel angle should be introduced, for example at 5° of heel.

There are now two possible equilibrium states for the boat, which require a rearrangement of the state and decision variables as the boat switches between them. In strong winds the boat is righting moment limited, that is the boat equilibrates to a fixed righting moment equal to sum of static righting moment at a prescribed heel limit plus righting moment from the crew at their maximum hiking or trapezing position. The sails are trimmed and twisted to equal this value and maximise speed. In light winds, in the pre-limit condition, the boat heels to some point less than the heel limit, sails are set at maximum lift coefficient (no trim or twist), with the skipper/crew righting moment determined as some function of the

138 Sailing Yacht Design: Theory

ratio of heel angle to the heel limit. In this pre-limit state the conventional arrangement of state and decision variables may be used but once in the righting moment limited condition heel angle, one of the decision variables, is fixed, while still having two equilibrium equations to satisfy. Thus, in order to solve for an equilibrium condition a new decision variable is used, the sail flat term, leaving the sail reef or twist parameter and the true wind angle as decision variables by which speed or VMG are optimised.

Other variables can be introduced into the optimisation procedure, such as centreboard extension, increasing the complexity of the optimisation, but more clearly reflecting the manner in which the boats are sailed. This type of optimisation routine is normally introduced outside the core equilibrium calculation as an additional iterative loop.

Catamaran

The catamaran presents a further extension of the unballasted sailing dinghy in that it has two states of sailing. In very light winds it will sail with both hulls immersed in the water, but as the wind strength increases the leeward hull will carry an increasing proportion of the boats displacement in order to provide a righting moment. Ultimately the windward hull will be lifted clear of the water, the wind strength when this occurs being dependent on the height of the sailplan and the displacement and beam of the catamaran. For light weight catamarans flying a hull is the preferred modus operandi as it represents the lowest drag configuration due to having a minimum wetted surface area.

The catamaran assumes two states of equilibrium, which are the fly and no-fly condition, as shown in Figure 7.12. While there are two 'states', the solution procedure considers three sets of circumstances, two of which are associated with the fly condition and the other with the no-fly condition.

There are two special heel angles that determine the boundaries of operation, firstly the 'break' angle, the heel angle at which the weather hull breaks or clears the free surface and secondly the fly-angle, which is the mean heel angle at which the boat will sail while flying the weather hull.

In the 'no-fly' condition the crew cannot derive from the sails, a heeling moment which exceeds the average of the righting moments associated with the fly-angle and break angle, with the crew assumed to be at centreline. The sails are trimmed and twisted for maximum lift. The boat equilibrates to some heel angle between zero and the break angle, with the crew moving from centreline to the farthest outboard position depending on the heel angle.

The next state is flying a hull with the crew at less than their maximum righting arm. The boat can achieve a heeling moment greater than the righting moment at the fly-angle, but less than the sum of the fly-angle righting moment and maximum crew righting moment. The sails are trimmed and twisted for maximum lift. The crew has moved outboard to produce the additional righting moment in order to balance the heeling moment. In this condition heel is replaced as a state variable by the crew righting moment expressed as fraction of the maximum available, leaving flat and twist, or reef as the decision variables available to maximise speed.

The final condition as wind speed increases further is flying a hull with the crew at their maximum righting arm position. The boat is thus righting moment limited at a value equal to sum of static righting moment at a prescribed fly-angle plus the righting moment from the crew at their maximum outboard position. The sails are trimmed and twisted to equal this value and maximise speed, in a similar way to the sailing dinghy.

One difficulty of producing robust performance predictions for these high-speed sailing yachts with low residuary resistance is that a relatively small increase in thrust from the sails produces a significant increase in sailing speed, and this produces an increase in apparent wind speed, and a further increase in speed. Thus in seeking optimum true wind angles the Newton Raphson optimisation cannot detect optimum sailing conditions that are beyond those close to its starting value, for example in such cases where a monohull might lift onto the plane and produce a higher Vmg, or the multi-hull might create enough apparent wind to achieve the flying state. These problems are most commonly overcome by programming 'tricks' based on a knowledge of expected behaviour.

7.7 PRESENTATION

The presentation of VPP results falls into two categories:

1. The printing and plotting of results for individual boats so that speeds and force audits can be quickly determined
2. The post processing of the results of a fleet of yachts to determine the 'best', however that might be defined

Single boat output

In a VPP calculation for a complex yacht such as a Whitbread 60 the calculations for a single yacht will contain results for several operational sets (OpSets) that define combinations of flotations and sailsets. The simplest form of output is the 'best' speed polar diagram, that is the envelope of the best speed produced by each opset. Figure 7.13 shows such a diagram, which defines the boat speed at each true wind angle, and also shows 'iso heel angle' curves which are useful in determining test matrices described in Chapter 16. The VPP will calculate the speed for each OpSet (combination of sailset and flotation) at each of the prescribed V_T and β_{TW} conditions and present the polar curve for the best speed. Performance is usually most easily assessed from tables of speed and associated sail set and flotation condition. Figure 7.14 shows tabular output for a notional yacht that carries water ballast and a reaching and running spinnaker. The table shows how the water ballast is taken on as the wind speed increases, and the asymmetric sail is used on reaching points of sail. More usual VPP output is shown in Figure 7.15, this is in three parts; a table of speed, heel angle, reef, and flat, followed by tables of hydrodynamic and aerodynamic force coefficients.

140 *Sailing Yacht Design: Theory*

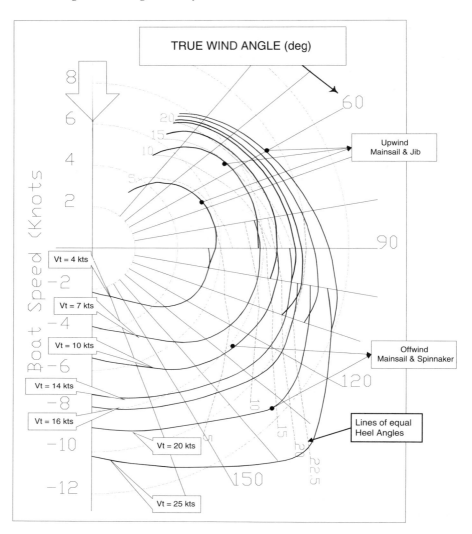

Figure 7.13 *Polar curve.*

Velocity Prediction Programs

Best Boatspeeds (kt)								Best Heel Angles (deg)						
TWA	True Wind Speed (kts)							True Wind Speed (kts)						
Deg	4	7	10	12	14	20	25	4	7	10	12	14	20	25
32	2.9	4.95	6.23	6.81	7.13	7.52	7.67	2.6	9.5	10.5	13.3	15	17.2	17.5
36	3.4	5.62	6.92	7.34	7.57	7.91	8.07	3	10.9	12.1	14.5	16.4	17.8	18.1
40	3.84	6.18	7.35	7.67	7.87	8.21	8.38	3.4	12.1	13	15.3	17.4	18.1	18.5
45	4.32	6.72	7.7	7.98	8.16	8.52	8.73	3.7	13.2	13.2	16.4	18	18.4	18.7
52	4.84	7.21	8.02	8.3	8.49	8.92	9.18	4.1	13.7	12.7	17.4	18.3	18.6	19
60	5.29	7.53	8.26	8.62	8.83	9.33	9.66	4.3	13	11.5	17.7	18.5	19.1	19.5
70	5.64	7.74	8.44	8.9	9.2	9.83	10.34	4.1	11.3	9.7	13.7	20.2	19.6	20
80	5.8	7.81	8.67	9.01	9.45	10.38	11.12	3.6	9.4	15.2	18	14.1	19.9	20.5
90	5.76	7.95	8.84	9.29	9.62	10.92	11.95	3.1	12.6	12.1	18.6	18.5	22.2	20.9
100	5.75	7.96	8.86	9.38	9.91	11.2	12.55	3.5	10.3	19.3	12.1	17.5	19.8	21.4
110	5.6	7.84	8.8	9.35	9.79	11.72	13.1	2.9	7.6	12.1	16.9	10.6	20.2	20.7
120	5.23	7.54	8.49	9.1	9.71	12.05	13.97	2.1	5.1	7.6	10.1	13.5	15.5	21.2
130	4.55	7.01	8.07	8.64	9.23	11.7	14.36	1.2	2.8	4.3	6	7.6	16.7	15.1
135	4.08	6.59	7.83	8.39	8.97	11.32	14.06	0.8	2	3.2	4.3	5.9	11.9	21.7
140	3.62	6.05	7.55	8.14	8.69	10.76	13.82	0.5	1.4	2.4	3.1	3.9	8.3	14.7
150	2.9	5.07	6.85	7.61	8.17	9.92	12.32	0.2	0.6	1.1	1.5	2	4.2	7.3
165	2.39	4.22	5.91	6.91	7.63	9.26	11.02	0.1	0.2	0.4	0.5	0.7	1.7	2.7
180	2.15	3.82	5.4	6.37	7.2	8.8	10.2	0	0	0	0	0	0.1	0.2
Best SailSet								Best Flotation						
32	Up	Up	Up	Up	Up	Up	Up	Base	Base	WB	WB	WB	WB	WB
36	Up	Up	Up	Up	Up	Up	Up	Base	Base	WB	WB	WB	WB	WB
40	Up	Up	Up	Up	Up	Up	Up	Base	Base	WB	WB	WB	WB	WB
45	Up	Up	Up	Up	Up	Up	Up	Base	Base	WB	WB	WB	WB	WB
52	Up	Up	Up	Up	Up	Up	Up	Base	Base	WB	WB	WB	WB	WB
60	Up	Up	Up	Up	Up	Up	Up	Base	Base	WB	WB	WB	WB	WB
70	Up	Up	Up	Up	Up	Up	Up	Base	Base	WB	WB	WB	WB	WB
80	Up	Up	RCH	RCH	Up	Up	Up	Base	Base	WB	WB	WB	WB	WB
90	Up	RCH	RCH	RCH	RCH	Up	Up	Base	Base	WB	WB	WB	WB	WB
100	RCH	RCH	RCH	RCH	RCH	RCH	Up	Base	Base	Base	WB	WB	WB	WB
110	RCH	RCH	RCH	RCH	RCH	RCH	RCH	Base	Base	Base	Base	WB	WB	WB
120	RCH	RCH	RCH	RCH	RCH	RCH	RCH	Base	Base	Base	Base	Base	WB	WB
130	RCH	RCH	RCH	RCH	RCH	RCH	RCH	Base	Base	Base	Base	Base	Base	WB
135	RCH	RCH	RCH	Dn	Dn	Dn	Dn	Base	Base	Base	Base	Base	Base	Base
140	RCH	RCH	Dn	Dn	Dn	Dn	Dn	Base	Base	Base	Base	Base	Base	Base
150	Dn	Dn	Dn	Dn	Dn	Dn	Dn	Base	Base	Base	Base	Base	Base	Base
165	Dn	Dn	Dn	Dn	Dn	Dn	Dn	Base	Base	Base	Base	Base	Base	Base
180	Dn	Dn	Dn	Dn	Dn	Dn	Dn	Base	Base	Base	Base	Base	Base	Base

= True Wind Angle
Up = Upwind SailSet. (Mainsail & Genoa)
RCH = Reaching SailSet (Mainsail & Gennaker)
Dn = Downwind SailSet (Mainsail & Spinnaker)

Base = Normal Flotation
WB = Water Ballasted Flotation

Figure 7.14 *Typical VPP output of speed, flotation and sail set for a yacht with multiple opsets (combination of flotation and sail sets).*

Performance Numbers Upwind OpSet

Vt	Bt	Vs	Vmg	Heel	Reef	Flat	Va	Ba	Leewy
10	32	6.504	5.515	16.8	1	0.774	15.78	18.7	3.02
10	36	6.988	5.653	18.2	1	0.763	16.03	20.3	2.71
10	40	7.322	5.609	19.2	1	0.763	16.11	22	2.53
10	45	7.62	5.388	20.1	1	0.778	16.07	24.3	2.37
10	60	8.189	4.094	21.2	0.986	0.912	15.41	31.4	2.09
10	70	8.465	2.895	21.7	1	0.997	14.69	36.1	1.94
10	80	8.647	1.501	16.7	1	1	14	42.2	1.55
10	90	8.637	0	11.8	1	1	13.04	48.5	1.23
10	100	8.45	-1.467	8.1	1	1	11.83	55.4	0.95
10	110	8.122	-2.778	5.2	1	1	10.47	63.3	0.71
10	120	7.647	-3.824	2.9	1	1	9.06	73	0.5

Vt = True Wind Speed Bt = True Wind Angle Vs = Boat Speed
Va = Apparent Wind Speed Ba = Apparent wind Angle

Hydrodynamic Force Numbers for Upwind OpSet

Vt	Bt	Rw/u	Rvc	Rva	Rh	Ri	Raw	Rt	Fh
10	32	0.158	0.426	0.208	0.006	0.201	0.176	82.8	457.3
10	36	0.214	0.413	0.2	0.01	0.163	0.136	99.7	479.5
10	40	0.281	0.387	0.187	0.013	0.133	0.108	116.9	494.2
10	45	0.355	0.351	0.169	0.018	0.108	0.083	138.2	506.2
10	60	0.498	0.273	0.131	0.029	0.069	0.039	198.2	526.6
10	70	0.553	0.242	0.116	0.035	0.054	0.022	234.8	528.4
10	80	0.602	0.239	0.111	0.015	0.033	0.011	250.7	459.1
10	90	0.616	0.249	0.113	0.002	0.021	0	244.1	371.2
10	100	0.597	0.273	0.122	-0.005	0.013	0	216.5	275.4
10	110	0.54	0.317	0.14	-0.005	0.008	0	174.9	185.9
10	120	0.432	0.392	0.172	0	0.004	0	127.5	113.5

Rw/u = Residuary resistance/Total Res. Rvc = Canoe body Viscous res.
Rva = Appendage Viscous resistance Rh = Drage due to heel
Ri = Induced drag Raw = Added Resistance in waves
Rt = Tot; Rt = Total Resistance Fh = Heeling Force

Aerodynamic Force Numbers @ CE for Upwind OpSet

Vt	Bt	ba	qan	cl	cdp	cdi	cdw	cdt	zce
10	32	18.55	482.12	0.954	0.024	0.089	0.026	0.139	7.39
10	36	20.14	497.76	0.973	0.024	0.094	0.026	0.143	7.36
10	40	21.84	502.98	0.999	0.025	0.099	0.026	0.15	7.35
10	45	24.07	500.23	1.037	0.027	0.108	0.026	0.161	7.34
10	60	31.1	460.27	1.202	0.044	0.153	0.026	0.222	7.24
10	70	35.82	418.55	1.352	0.067	0.191	0.026	0.284	7.33
10	80	41.76	379.91	1.341	0.101	0.185	0.026	0.313	7.32
10	90	48.04	329.53	1.304	0.142	0.175	0.026	0.342	7.32
10	100	54.85	270.99	1.238	0.185	0.16	0.026	0.371	7.32
10	110	62.64	211.86	1.136	0.236	0.138	0.026	0.4	7.34
10	120	72.26	158.13	0.987	0.304	0.108	0.026	0.438	7.39

qan = dynamic head * Sail Area Cl = lift coefficient
Cdp = Viscous drag coefficient Cdi = Induced drag coefficient
Cdw = Windage drag coefficient Cdt = Total drag coeffcient
zce = Height of centre of effort abose baseline

Figure 7.15 *Typical VPP performance and force audit output.*

Comparative analysis

Comparative analysis for a fleet of candidate designs can be undertaken at a range of different levels of sophistication, from a simple speed difference expressed as seconds per mile sailed, through to simulated regattas sailed over pre-determined courses using game theory, as described in reference 29. These approaches are described in *Sailing Yacht Design: Practice* – Chapter 8.

REFERENCES

1. Davidson, K.S.M, *Some Experimental Studies of the Sailing Yacht*, Transactions of The Society of Naval Architects and Marine Engineers, Vol. 44, 1936.
2. Kerwin, J.E., Newman, J.N., *A Summary of the H. Irving Pratt Ocean Race Handicapping Project*, 4th Chesapeake Sailing Yacht Symposium, Oct., 1979.
3. Hazen, G.S., *A model of sail aerodynamics for diverse rig types*, Proceedings of New England Sailing Yacht Symposium, 1980.
4. Van Oossanen, P., *Predicting the Speed of Sailing Yachts*, Transactions of the Society of Naval Architects and Marine Engineers, Vol. 101, 1993.
5. Claughton, A.R. and Oliver J.C. *Development of a multi-functional velocity prediction program (VPP) for sailing yachts*. International Conference CADAP95, RINA 1995.
6. Schwenn, P., Hazen, G., *Drawing with Performance Prediction,* Proceedings of the 12th Chesapeake Sailing Yacht Symposium, Jan., 1995.
7. Stephens O.J., *Guides to the approximation of sailing yacht performance*, 9th CSYS, SNAME, 1989.
8. Keuning, J.A., Sonnenberg, U.B., *Developments in the Velocity Prediction based on the Delft Systematic Yacht Hull Series*, The Modern Yacht Conference March 1998.
9. Henriksen, K., Jensen, P.S., *Velocity Predictions for Yachts at the Danish Maritime Institute*, Report of the Danish Maritime Institute, Sept. 1989.
10. Oliver, J.C., *Performance Prediction Method for Multihull Yachts.* Proceedings of the 9th CSYS, SNAME., 1989.
11. Teeters, J.R., *Refinement in the Technique of Tank Testing Sailing Yachts and the Processing of Test Data*, Proceedings of the 11th Chesapeake Sailing Yacht Symposium, Jan., 1993.
12. Hoerner, S.F, *Fluid dynamic drag*, Hoerner 1965.
13. Poor, C.L., *A Description of the New International Rating System*, Publication of the United States Yacht Racing Union (USYRU), 1986
14. Claughton, A. *The effect of counter length on hull resistance*, 9th CSYS, SNAME, 1989.
15. Keunung, J.A., Binkhorst, B-J., *Appendage resistance of a sailing yacht hull.*, Transaction of 13th CSYS, 1997.
16. Schlageter, E.C., Teeters, J.R., *Performance Prediction software for IACC Yachts*, Proceedings of the 11th Chesapeake Sailing Yacht Symposium, Jan., 1993.
17. Marchaj, C.A., *Aero-Hydrodynamics of Sailing*, revised edition, Adlard Coles, London, 1988.
18. Tinoco, E.N. *et al, IACC Appendage studies.* Proceedings of 11th CSYS, SNAME, 1993.
19. Nomoto, K., *Balance of helm of a Sailing Yacht*, Proceedings of 6th HISWA Symposium on Yacht design and Construction, Amsterdam, Nov. 1979.

20. Sclavounos, P.D., Nakos, D.E., *Seakeeping and Added Resistance of IACC Yachts by a Three-Dimensional Panel Method*, Proceedings of the 11th Chesapeake Sailing Yacht Symposium, Jan., 1993.
21. McRae, B., Binns, J., Klaka, K., *Windward Performance of the AME CRC Systematic Yacht Series*, The Modern Yacht Conference, RINA, March 1998.
22. Milgram, J.K., *Fluid Mechanics for Sailing Vessel Design*, Annual Review of Fluid Mechanics, 1998.
23. Drela, M., *An analysis and design system for low Reynolds number aerofoils.*, In Lecture Notes in Engineering Low Reynolds Number Aerodynamics, Vol 54, Springer Verlag.
24. Larsson, L., *Numerical predictions of the flow and resistance components of sailing yachts.* Proceedings from Conference on Yachting Technology, University of West Australia, Perth, 1987.
25. IMS Rule Book. Offshore Racing Council.
26. Glauert, H., *Airfoil and air screw theory*, Cambridge University Press, 1959.
27. Claughton, A.R. and Campbell, I.M.C., *Wind Tunnel Testing of Sailing Rigs,* 13th International Symposium on Yacht Design and Construction (HISWA), 1994.
28. Milgram, J.K., *Naval architecture technology used in winning the 1992 America's Cup match.* Transactions Society of Naval Architects and Marine Engineers (SNAME), 1993.
29. Oliver, J.C., Letcher, J.S. and Salvesen, N., *Performance Predictions for Stars and Stripes*, SNAME Annual meeting, 1987.

CHAPTER 8
MATERIALS IN CONSTRUCTION

R. Loscombe and A. Shenoi*
Southampton Institute and University of Southampton*

8.1 PARAMETERS INFLUENCING CHOICE

General

In order for any material to be a viable candidate for use in sailing yacht construction, it must be able to demonstrate adequate performance across a range of requirements. The principal requirements are:

- *Adequate strength*
    - Under 'once in a lifetime' loading ('**ultimate**' strength)
    - Under long-term static loading (**creep** resistance)
    - Under long-term dynamic loading (**fatigue** life)
- *Adequate stiffness*
    - Under lateral loading (panel flexural rigidity)
    - Under inplane loading (**buckling** resistance)
- *Ease of fabrication*
    - Ease of forming
    - High joint efficiencies
- *Fire resistance*
    - Fire containment
    - Minimum toxicity levels
- *Impact resistance and damage tolerance*
    - 'Minor' damage absorption capability
    - 'Major' damage absorption capability
- *Long-term performance of material*
    - Low material wastage
    - Acceptable degradation of mechanical properties with time
- *Cost*
    - Raw materials
    - Associated tooling, capital, labour

Strength considerations

Strength is the ability to carry load at a defined level of 'damage'. A sailing yacht structure must be designed for a range of load cases. For example, suppose a typical cruising yacht is at sea for 10% of its anticipated lifetime of 30 years, i.e. about eight hours per day, every weekend for nine months of the year. The probability of encountering the extreme 'once in a lifetime' load in this period is about 1×10^{-7}. This extreme load could occur on day one or day 11,000 and so a prudent designer would need to use the long-term mechanical properties; the short-term properties obtained from mechanical testing need to be derated by an appropriate factor. As this is an extreme load, some damage may be permitted. Therefore, the **ultimate strength** is one key property.

However, it should not be inferred from this that the extreme load may always be equated with a capability based on the mean ultimate strength. This is only possible when the extreme load is known to great accuracy and even then the designer may wish to impose further requirements such as 'once in one-hundred boat lifetimes'. In practical design, the level of uncertainty associated with loads and material properties means that capability may have to be based on a lower estimate of strength. This is the area of partial safety factors which is discussed in Chapter 15. The variation in material property as measured by the coefficient of variation (COV) will have a significant effect on the value of the partial safety factor.

The yacht will also be subjected to lesser loads for longer durations. In order not to **accumulate damage**, it is important that the limit of proportionality strength should be as high as possible. For metals, this is traditionally taken as close to the yield strength (steel) or 0.2% strain offset strength (aluminium alloys). Both wood and composites (particularly composites under tension) exhibit a 'knee' in the stress-strain curve. This typically occurs at 30–60% of the ultimate strength.[1,2] This proportional limit or **elastic limit** is a second key property. As in the case of extreme load case design, uncertainties in the process mean that factors of safety (traditional or partial) will usually be required.

Certain components are subjected to long-term static or 'dead' loads; for example engine seats. If these loads cause stresses beyond a certain level, strains will continue to increase leading to failure. This tendency is represented by the **creep limit**. This is defined here as the stress (or percentage of ultimate strength) below which creeping may be neglected. A figure of 20% is typical for composites.

Other components are subjected to harmonic or near harmonic dynamic loads; for example masts, engine seats, keel structure. This lifetime fatigue loading may be represented by n stress groups where each group consists of a stress range ($\sigma_{max}-\sigma_{min}$), maximum stress and number of cycles experienced. The fatigue resistance of a material may be characterised by $S-N$ data, where S is the stress amplitude (or range) required to cause failure (usually fracture) at N cycles. The fatigue life may be estimated using a damage accumulation rule such as Miners. One simple measure of resistance to fatigue is the **fatigue limit**; that stress at which the $S-N$ curve becomes horizontal. Many materials do not exhibit an exact fatigue limit, but this is traditionally taken as the S value (where $N = 10^8$).

It should be noted that for sailing yachts, the endurance limit at fewer cycles may be more relevant. A private cruising yacht case may well experience a lot less than 10^7 cycles, as would a competition racing yacht intended for a lifespan of less than a few years.

Stiffness considerations

Stiffness is a measure of the resistance to deformation under load. In particular, the decks of sailing yachts must resist lateral loads such as crew movements on deck, and the destabilising inplane loads due to the sagging moment produced by the standing rigging. The tensile modulus (i.e. Young's modulus) is often used as an indicator of stiffness. However, material rankings based on the ratio of modulus to density can be misleading since stiffness is also a function of geometry. Excessive stiffness is undesirable for impact loaded panels. Firstly, the panel will not be able to utilise membrane forces before the failure stress is reached. Secondly, the ability to absorb energy may be unduly limited.

Fabrication considerations

Although it is possible to develop conic hull forms for a sailing yacht hull, like any other, most hulls will have compound curvature and so the ability to form becomes a major consideration. The relative ease with which composite or wood hulls can cope with such curvature is one of the many reasons for the attractiveness of in-situ laminated construction methods.

Metal structures are still predominantly welded. The key problems associated with small craft are related to avoiding distortion, burn-through and residual stresses since thicknesses tend to be quite low.

Adhesive bonding may be used when making secondary bonds or joining dissimilar materials. Many sailing yachts employ composite hulls with plywood bulkheads, but adhesive bonding offers the possibility of joining metals to wood and composite. Epoxy and urethane acrylate based resins are available.[3] Such joints should be designed so as to minimise peel loads.[4] (Figure 8.1.)

It is important that the fabrication process does not produce large spatial variations in properties, for example due to careless orientation of fibres or poor control of resin:fibre ratio. Design calculations are often based on mean property less two standard deviations. Therefore, a structure having high COVs will end up heavier than a similar boat having the same mean property but with a smaller spread. This is largely a matter of quality assurance, although wood is more difficult to control being a wholly natural material.

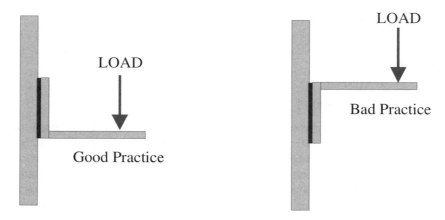

Figure 8.1 *Adhesive bonding of joints.*

Fire resistance considerations

Any structural material intended to provide some degree of fire resistance must be able to satisfy short-term performance criteria; ease of ignition, flame spread, production of smoke and toxic gases and heat generation, and long-term criteria involving residual strength and stiffness.[5] Reinforced plastics conduct heat poorly, unlike metals, but have other problems associated with toxicity and smoke obscuration. Epoxy and polyester resins do not char and hence produce considerably more smoke than phenolic resins.[4]

Impact considerations

Yacht structures must be able to resist collision with floating objects and groundings, in addition to slam or wave surfing impacts. A material may be characterised by its ability to absorb energy and must be adequately **notch tough** to the inevitable surface damage likely to result. Toughness is also a function of operating temperature. A yacht intended for arctic service would need to employ a material which was not liable to brittle fracture.

Impact damage may be characterised as minor and major. Minor impact energies, such as may occur in low-speed manoeuvring, should be absorbed elastically (i.e. without permanent damage). The strain energy per unit volume (resilience) at the elastic limit is $0.5\sigma_e^2/E$. For major impacts, permanent damage is acceptable subject to the vessel remaining afloat. The corresponding resilience being:

$$\int \sigma(\varepsilon)\, d\varepsilon$$

Typical resilience values:

- *At the elastic limit*
  - 0.19 MPa (aluminium)
  - 0.13 MPa (mild steel)
  - 0.13 MPa (GRP roving)
- *At the failure limit*
  - 24 MPa (aluminum)
  - 80 MPa (mild steel)
  - 1.4 MPa (GRP)

These formulations assume that full direct strength may be exhausted in absorbing energy. For composites, failure between plies or shearing may occur before this happens. Strain rate effects may also lead to a modification of mechanical properties.

Long-term considerations

While steel may corrode and lose thickness, the mechanical properties are not seriously denuded merely by the passage of time. For composites, some mechanical properties will reduce with long-term immersion. Water penetrates by diffusion through the resin and capillary flow through cracks and voids. Mechanical properties are degraded partly due to damage to the fibre resin bond (resin swelling) and chemical attack. Smith[6] suggests FRP short-term properties should be reduced by 20% to take account of degradation in a wet environment.

Cost considerations

Comparing costs for alternative materials is exceedingly difficult, since this largely depends on the starting point. A yard specialising in aluminum alloy may find the cost of retooling/retraining to build in composites to be prohibitive. Design costs may also be an issue. Designing an aluminium alloy hull to class rules may be a fraction of the cost of a fully engineered advanced composites solution, albeit that the latter results in a 'better' structure. Alternatively an independent designer/specialist structural design bureau may be able to approach a range of yards and so this issue may not arise. Material costs vary enormously with E–glass-polyester composites ranging from about £3/kg up to about six times as much for unidirectional epoxy prepregs.[7] Carbon prepreg skins with Nomex honeycomb core would be about ten times the cost of conventional GRP single skin.[8] Steel would be about half the cost per kg of conventional GRP. Aluminium sheet would be about the same as GRP.[7]

It is generally accepted that comparisons based on raw material cost have limited value. For production craft, it is the overall cost per unit that matters. For racing yachts, cost may be virtually irrelevant in the drive towards the minimum weight solution in order to maximise ballast ratio and hence sail area capacity. One simple objective function which enables cost to be included is of the form;

150 *Sailing Yacht Design: Theory*

$$U = \alpha \frac{W}{W_0} + (1-\alpha)\frac{C}{C_0} \qquad [8.1]$$

U = the function to be minimised
W, C = the structural weight and fabricated cost of a given design
W_0, C_0 = the weight and cost of a standard or reference design
α = weighting factor which varies between 1 (minimum weight objective) to 0 (minimum cost objective)

The cost may be crudely estimated at the preliminary design stage using:

$$F_C = O_F \cdot H_W \cdot (1 + S)\,[M_R + L_P \cdot W_R + T_C + M] \qquad [8.2]$$

F_C = fabrication cost
O_F = overhead factor
H_W = hull weight in kg
S = proportion of scrap material
M_R = material rate in £/kg
L_P = labour productivity in manhours/kg
W_R = wage rate in £/hour
T_C = tooling cost per kg worked
M = margin

The tooling cost could include items such as the release film, peel ply, vacuum bag film, wood for moulds etc.

8.2 STEEL

Characteristics

Steel, the traditional choice for many commercial boats, has also been used for sailing yachts. The density and Young's modulus values are about three times those of aluminium alloy. The yield strength is about twice that of aluminium. It is a robust, cheap, easily weldable material with a modulus greater than any other boat-building material with the possible exception of certain types of carbon composite. The fire resistance of steel is often taken as the standard against which other materials are required to show equivalence. The extent to which steel has been used for sailing yachts is limited by weight and maintenance considerations.

Steel types

Steels may be broadly divided into three classes:

- Normal mild steel (MS): Typical minimum yield stress 235 MPa[9]
- Higher strength steels (HTS): Typical yield stress range 265–390 MPa
- High yield steels (HY): Typical yield stress greater than 400 MPa

These may be further classed by notch toughness, which involves defining the impact toughness test temperature, typically from 0 °C to about –60 °C. The Young's modulus and density do not vary very much between steel types being about 210 GPa and 7800 kg/m^3 respectively. The ultimate strength:yield strength ratio varies from about 1.3 to 2, indicating a considerable post yield capability. Steels exhibit a distinct fatigue limit after about 10^7 cycles. Reference 10 quotes a stress range of 100 MPa for class B (basically unwelded) and 47 MPa for class E (manual transverse butt weld). These figures are based on mean minus two standard deviations S–N data. However, for fillet welded surface mounted lugs and similar, the allowable stress range can be down to 30 MPa. Readers are cautioned to clarify whether a stress range or amplitude is being quoted and to always refer to the source reference.

Construction considerations

A typical plating thickness for aluminium alloy would be 4–8 mm. In order to have the same shell weight, a steel panel would need to be 1.5–3 mm thick. Welding such thin plating is difficult. In addition, unprotected mild steel corrodes at the rate of about 0.05 mm[11] per year and such thin skins might be thought to have an inadequate corrosion margin. Low alloy corrosion resistant steel containing traces of copper such as Cor-ten corrode at less than 0.01 mm per year in industrial atmospheres but may be only a little better than mild steel in marine conditions.[11]

Welding of high yield steels requires careful control of pre-heat and post-weld cooling. High strength low alloy (HSLA) steels offer similar strength to HY80 but are said to be cheaper and easier to weld.[12] The use of high and higher strength steels in small craft is limited by the fact that scantlings are probably down to practical minimums. In addition, there may be no gain in buckling resistance (Young's modulus dependent) and crack initiation resistance (weld quality dependent) from that provided by mild steel.

Steel is a strength limited material, that is scantlings are governed by strength. There is very little possibility of utilising membrane resistance, leading to the inevitable structural arrangement of closely framed stringers for minimum weight designs.

8.3 ALUMINIUM ALLOYS

Characteristics

Aluminium alloys are typified by a specific gravity of 2.6–2.8 and a Young's modulus of about one third of steel, i.e. 69 GPa. They possess good corrosion resistance due to a self-sealing, nearly invisible, protective oxide film which forms naturally on the surface immediately on exposure to air. Some alloys may acquire a grey, pebble like appearance in some marine environments and may become darker and rougher in moderate industrial atmospheres. However, tropical climates are no more detrimental than temperate ones.[13]

While requiring more care than steel, most aluminium alloys are readily weldable using TIG (tungsten inert gas), MIG (metal inert gas) and pulse-arc. TIG is slow compared with the other processes and is associated with high heat build-up. The speed of operation does allow the welder to produce high quality welds and TIG is often used where the weld bead is intended to be the finished product. MIG is a fast semi-automated process which is used for 3 mm plating and thicker. For thinner plating it is possible to use pulse-arc where pulsing on and off limits the excessive weld heat build-up which can occur when welding very thin sheets using MIG.[14] The main difficulties associated with aluminium alloy concern the effect welding has on the fatigue life and static strength and the different thermal characteristics. Adhesive bonding offers an alternative to riveting for very thin sheeting and may eventually overcome some weld associated problems.

Alloy types

Aluminium alloys used in yacht construction are usually the non-heat treatable magnesium alloys or heat treatable magnesium–silicon forms. Non-heat treatable alloys are those for which strength is increased by strain-hardening, whereas heat treatable derive increased strength by one or two stages of heat treatment. The first phase is called solution heat treatment and consists of heating to a high temperature and quenching. The second phase of precipitation heat treatment or ageing is where the alloy is maintained at a moderate temperature for some time. For some alloys, ageing occurs naturally and so the second phase may not be required.

There are a number of ways of classifying aluminium alloys. Det norske Veritas[9] and the International Organisation for Standardisation nomenclature indicate the main alloying elements and the percentage. For example, 'AlMg4,5Mn' denotes a non-heat treatable alloy, with 4.0–4.9% magnesium and 0.3–1.0% manganese. An alternative system adopted by the Aluminium Association, Lloyd's Register of Shipping[15] and BS 8118[13] uses a four digit designation and a temper key. This form is preferred here since it is felt to be more memorable.

The first digit denotes the major alloying element; 1 corresponds to 99.99% pure aluminium, 5 to magnesium, 6 to magnesium and silicon, 7 to zinc. Hence 5000 series denotes non-heat treatable alloys and 6000 series denotes heat treatable. The second digit indicates alloy modifications with 0 denoting the original alloy. The last two digits are of no special significance other than they allow different alloys to be identified.[13] Table 8.1 contains details of the most common alloys used in marine construction.

The mechanical properties are also dependent on the condition or temper of the alloy. Common nomenclature employs O for the fully annealed condition and F for as fabricated, meaning some temper is acquired from shaping and no special control has been maintained over thermal conditions. H denotes strain-hardened with subsequent digits indicating the final degree of strain hardening, that is H1 refers to strain-hardened only, H2 to strain-hardened and partially annealed and H3 to strain-hardened and stabilized (i.e. subjected to a final low temperature heating process). T denotes solution heat treatment with the subsequent digit indicating the post phase one treatment.

Table 8.1 Princilap marine grade aluminium alloys

Four Digit Designation	ISO Designation	Silicon (%)	Magnesium (%)	Manganese (%)
5086	AlMg4	0.5 max.	3.5–4.6	0.8 max.
5083	AlMg4.5,Mn	0.5 max.	4.0–4.9	0.3–1.0
6082	AlMgSil	0.6–1.6	0.4–1.4	0.4–1.0

Compositions taken from reference 9, which are slightly different from reference 15 since the designations are not identical.

The majority of marine grade alloys will probably be drawn from the following list; 5086 (O/F/H321), 5083 (O/F/H321), 5454, 6061–T6, 6082–T6. Most of these alloys are available in sheet, plate and extrusion form. There are important differences which will affect the final choice. For example, 6061 is not normally acceptable for application in direct contact with sea water[15] but exhibits a slight improvement in extrudability and surface finish over 6082.[13] Hence, it might be used to best effect in complex, visible deckhouse extrusions. 5454 offers excellent resistance to corrosive attack especially in a marine atmosphere.[13] There are also custom and practice differences where 5083 is used in preference to 5086 in the UK and vice-versa in the USA.

The two most common alloys are 5083–O (N8 in early UK notation) and 6082–T6 (H30). Table 8.2 compares these alloys.

Table 8.2 Comparison of properties: 5083 v 6082

Alloy	Unwelded[a] 0.2% proof stress (MPa)	Unwelded[a] tensile strength (MPa)	Approximate loss of strength on welding[b]	Durability rating[c]
5083–O	125	275	0	A
6082–T6	240	295	50	B

[a] These are indicative figures from reference 13 for sheet, plate and extruded forms.

[b] Applies to bending and overall yielding. 45% for shear, tension and squash modes.

[c] A class requires protection only in severe marine atmospheres. B class means protection is required for all atmospheres and immersion in salt and fresh water.

[d] Values apply for typical yacht hull thickness.

Reference 13 quotes a stress range of 42 MPa (these stresses are mean – two standard deviation values) for detail class 60 (non-welded detail) down to less than 20 MPa for fillet welded lugs on the surface. Deck fittings are commonplace on yachts, so this diminution of fatigue strength is particularly relevant.

Construction considerations

As with steel, lightweight aluminium hulls tend to employ a closely spaced stringer form of construction. Conventional arrangements use either closely framed transverses with girders or closely framed longitudinals on widely spaced web frames. Although hull girder stiffness is less of a problem for aluminium craft, longitudinal framing offers some fabrication advantages. Allday[16] makes the point that improving the buckling resistance by extensive bracketing of end connections produces a complex weld zone which reduces the fatigue life. Here is an example of the difficulty of balancing the demands imposed on a structure.

Material options include:

- 5000 series plate and stiffeners
- 5000 plate and extruded 6000 stiffeners
- 6000 series extruded planks

Most joints are welded but use of adhesives is increasing, resulting in a return to the type of stiffener profile associated with riveted construction (i.e. the 'Z' stiffener). Welding of 6000 series causes a reduction in strength for about 25 mm from the joint. In a plate-stiffener combination, the HAZ lies close to the neutral axis and hence the loss in strength may be of no consequence. Extruded planks remove the weld line to a point of *potentially* low stress (Figure 8.2).

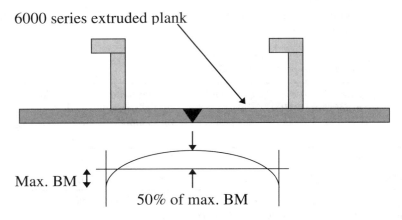

Figure 8.2 *Locating weld lines in low stress regions*

Figure 8.2 corresponds to a uniformly distributed load over regularly spaced stiffeners (i.e. 100% fixity). However, if the panel was subjected to a patch load, boundary conditions would approximate to 0% fixity, with the maximum moment occurring at mid span. Perhaps a safer course of action is to place the weld off centre, say at 25% of the span in from the stiffener. This corresponds to a low stress region irrespective of end fixity and will allow 6000 series to be designed to the unwelded strength providing that the loss of strength due to welding does not exceed about 50%, subject of classification society approval. The extent to which such extruded planks are suitable for use in sailing yachts is debatable, owing to hull curvature.

8.4 WOOD

Characteristics

Wood is a material which has densities in the region of 300–800 kg/m^3 and has the lowest density of all boat-building materials with the exception of sandwich cores. It is a highly orthotropic material with perpendicular to grain properties being 3–8% of corresponding values parallel to grain. Tensile moduli (parallel) are typically in the range 8–14 GPa and hence similar to conventional glass reinforced plastic.[17] The material axis are very nearly cylindrical in a plane perpendicular to longitudinal direction owing to the formation of the growth rings. This means the radial axis are not parallel but diverge. For laminated construction, veneers are thin enough for this to be neglected and wood material properties are usually referenced to three mutually perpendicular axes (L, T, R) Figure 8.3. The properties in R and T directions are similar but not identical. The plane of rotary peeled veneers as used in plywood sheets would be the L–T plane, whereas for quarter sliced veneers, this is the L–R plane. However, vagaries between plain sawn and quarter sawn blur the distinction. See reference 18 for a more detailed discussion.

156 *Sailing Yacht Design: Theory*

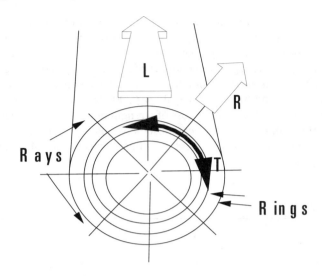

Figure 8.3 *Principal axes used to define material properties of wood.*

Traditional methods of carvel or clinker on frames have largely been superseded by laminated methods of construction. The veneers are 'encapsulated' in resin (usually epoxy) and may be constructed in a number of ways as described in the section on construction considerations. The net effect is to produce a stable structure, albeit with some increase in apparent density.

Wood types

Wood may be classified broadly by species, density (usually defined as oven dry mass divided by volume at actual moisture content) and moisture content. A typical moisture content would be about 6–20%. At about 30% moisture content, wood is considered to be in the 'green' condition and this is often taken as the upper practical limit. Material properties generally increase as density increases and moisture content reduces. A figure of 12% is typically taken as the reference value when quoting properties.

In order to characterise wood it is necessary to have tensile and compressive strengths parallel and perpendicular to the grain, inplane shear strength and modulus and elastic moduli in two directions as a minimum. This can result in an unmanageably large test programme. Properties do vary between species of nominally the same density, but the inherent variation between samples of the same species are likely to be as significant. Consequently, it is often sufficient at the initial design stage to specify hardwood or softwood and a suitable density and use regression equations.[17] For example:

$$E_L = 21.6 \, SG^{0.9} \text{ GPa} \quad \text{softwood 12\% moisture content} \quad [8.3]$$

This equation gives 11.2 GPa for douglas fir (*SG* 0.48) and 7.8 GPa for western red cedar (*SG* 0.32). These may be compared with values quoted in reference 20 of 13.5 and 7.7 GPa. An alternative approach is to determine the tensile strength or modulus of rupture and the elastic modulus by test and use previous data for the other properties, for example as laid out in British Standard EN 384. The modulus of rupture is the equivalent stress in the extreme fibres of a specimen at the point of rupture assuming that bending theory applies.[19] For three point bending:

$MOR = 3PL/2bd^2$

P = failure load
L = span
bd = cross-section of the specimen

The term *MOR* is generally only used for the parallel to the grain strength.

Some wood data apply to structural size samples, typically 50 mm × 100 mm cross-section (applicable to bent or steamed frames) and other data to clear, straight-grained specimen size coupons which are more applicable to veneers. The former contain defects such as knots while the latter do not. For structural size samples, knots act as stress concentrators and the tensile strength is only about 60% of the MOR and similar to the compressive strength. For coupons, the parallel tensile strength is much greater than MOR and the compressive strength is much lower. This shows that great care is needed in using published data for wood.

Coefficients of variation on measured properties for wood are typically 20% (softwood) and 10% (hardwood).[17, 19] This is certainly greater than variations expected in steel or aluminum alloy and probably worse than a carefully controlled composite layup. It is normal structural design practice to work with characteristic properties (i.e. the 5 percentile value) and therefore wood will suffer to a greater extent than most other boat-building materials.

Construction considerations

In-situ moulded, laminated plywood hulls may be constructed in numerous ways:[20]

- Three or more veneers laid over closely spaced stringers, with lightweight glass cloth protection
- Strip planked inner skin over more widely spaced frames, skinned with wood veneers or composite
- Lightweight core (e.g. end grain balsa) faced with wood veneers (e.g. Duracore™)

It is extremely easy to produce a cold moulded hull laminate which is very unbalanced and may require a more sophisticated analysis than is generally used in small craft structural design.

8.5 COMPOSITES

Characteristics

The first point to be made is that this chapter cannot do justice to this topic, such are the large number of variables involved. Composites have densities which are typically 45–70% and Young's modulus some 10–200% those of aluminium alloy, with similar strengths. The material may vary from isotropic to highly orthotropic. Composites are composed of stiff fibres embedded in a flexible, tough resin matrix. The fibres carry the bulk of the loads whereas the resin protects the fibre and allows load transfer.

Composite types

A composite skin may be classified into single skin, sometimes called monolithic, or sandwich. Dealing with the monolithic or sandwich face, a laminate may be specified by the stacking sequence and ply list.

Stacking sequence (orientation of the warp direction for each ply)

For example, $[90/0/90]_s$ denotes a cross-ply laminate which is symmetrical about the mid-plane. While it is generally advantageous to place the stiffer fibres at maximum distance from the neutral axis, there are occasions when this does not allow the full strength of a laminate to be utilised.

Ply specification

- *Fibre types*
    - Carbon (HS – high strength; HM – high modulus; IM – intermediate modulus)
    - Aramid
    - Glass (E, S/R)
    - Polyester
    - Intraply hybrid (e.g. carbon warp, aramid weft)
- *Form*
    - Chopped strand mat (CSM)
    - Woven (plain, twill, satin)
    - Multi-axial
    - Unidirectional
- *Balance or bias*
    - Per cent fibres in warp direction
    - Per cent fibres in weft direction
- *Finish*
    - Resin compatible
- *Areal density*
    - Typically 200–800 g/m^2 or higher

- *Resin*
  - Polyester
  - Vinylester
  - Epoxy
  - Phenolic
  - Polypropylene
- *Fibre:resin fraction*
  - Fibre volume fraction (V_f) = volume of fibre:volume of laminate
  - Fibre weight fraction (W_f) = weight of fibre:weight of laminate

Core specification

- *Type*
  - Foam (linear or cross-linked)
  - End-grain balsa
  - Honeycomb
  - Aluminium (Aeroweb™)
  - Aramid coated paper (Nomex™)
- *Density*
  - Typically 50–300 kg/m^3
- *Thickness*
  - Typically 20–80 mm

Ply properties are proportional to the fibre volume fraction. For example, the Young's modulus is given by:

$$E = \alpha E_f V_f + E_R(1 - V_f) \quad \text{assuming no voids} \tag{8.4}$$

E_f, E_R = modulus of fibre and resin
α = fibre orientation related efficiency factor

E–glass based, CSM/WR layups have E values of about 10–13 GPa (i.e. similar to wood), whereas unidirectional carbon layups might be expected to exceed aluminium alloy values. The term advanced (previously exotic) composites usually implies the use of carbon and/or aramid. It should be noted that there is often a synergic effect when using hybrids.[6]

Construction considerations

The fibre volume function is related to the production method, with low values associated with wet layup, manual consolidation methods (e.g. 20% for CSM, 30% for woven roving glass) and higher values associated with pre-preg, vacuum bagging methods (e.g. 50%+). Aramid based laminates exhibit compressive strengths which are often no better than glass and it is therefore advisable to locate such fibres in tensile zones. High performance sailing yachts have been known to use an aramid/carbon hybrid in the outer faces of sandwich panels and all carbon in the inner skin. Under a pressure loading the outer skin will be in compression midway between frames, but be in tension at the frame. (See Figure 8.4.)

For 100% fixity, the ratio of maximum applied stress to stress at centre will be 2. Providing the compressive strength is greater than 50% of the tensile strength, the design works. This will not be the case for other fixities. In addition, it is important to adjust the thickness of the stiffer all carbon inner skin so that the neutral axis of the sandwich does not move so as to unduly 'load' the less stiff and often less strong outer face.

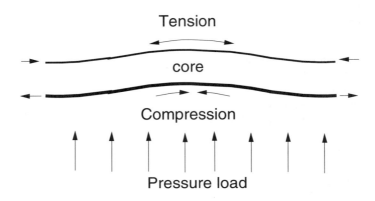

Figure 8.4 *Panel under pressure load.*

As with laminated wood construction, it is important for designers to fully appreciate the consequences of specifying unbalanced layups and the attendant problems associated with coupling of in-plane strains with out of plane forces and vice versa. (See chapter on Mechanics of Composite Materials.)

Core choice is often dictated by cost and function. Bottom panels subjected to high slam pressures will normally need to use a linear, high strain to failure core. Maximum stiffness is often the requirement for decks, where end grain balsa offers high shear modulus and wrinkling resistance at low cost. Weight efficient sandwich hulls tend to employ the thinnest faces needed to provide adequate robustness and the core depth and density are adjusted to provide the required shear strength. Since core properties are proportional to core density, the designer needs to balance the relative merits of a dense thin core against a low density thick core. This is not just a matter of weight and cost, but also the ease of fabrication, since moulding very thick cores is not without its problems.

8.6 CONCLUSIONS

There are occasions when material choice is dictated by client preference or limited by racing rule or national restrictions. Equally, there are occasions when the designer is free to select the best material for a particular application. Since there is no such thing as the ideal boat-building material, the selection procedure must weigh the advantages/disadvantages of a given material for a particular application using rational criteria.

It is a simple matter to calculate relative areal densities. For single skin panels, subjected to the same design pressure, with the same frame spacing, aspect ratio, deflection:span ratio and boundary conditions, the areal density of material **A** to material *B* is given by:

$$\frac{mass_A}{mass_B} = \frac{SG_A}{SG_B}\sqrt{\frac{\sigma_B}{\sigma_A}} \quad \text{strength} \tag{8.5}$$

$$\frac{mass_A}{mass_B} = \frac{SG_A}{SG_B}\sqrt[3]{\frac{E_B}{E_A}} \quad \text{stiffness} \tag{8.6}$$

SG = specific gravity
σ = design stress which takes into account the material dependent factor of safety
E = Young's modulus

Such equations may be used to indicate, for example, that a steel panel will be about twice the weight of an aluminum alloy one and that aluminium and GRP are about the same. Reference 8 indicates that conventional GRP will be about 5–20% heavier than aluminium alloy and an advanced carbon sandwich panel would be about 40% lighter than aluminium. These equations are useful for 'back of envelope', indicative calculations.

However, the temptation to provide further weight comparisons using such equations has been resisted here as the results can often paint a false picture. Instead, this chapter has attempted to identify some of the key mechanical properties, indicate how these influence their use in construction and hence demonstrate the complexity of the material selection process.

REFERENCES

1. Kollman, F.F.P. and Côté, W.A., *Principles of Wood Science and Technology - I Solid Wood*. George Allen and Unwin Ltd, London, 1968.
2. Niederstadt, G., *Background to Materials Science*. Vol 2, Chp. 2 - Composite Materials in Maritime Structures. Cambridge University press. 1993. ISBN 0 521 451531.
3. *High Performance Materials for Bonding Composite Structures*. Ship & Boat International, December 1995, pp29-30.
4. Mayer, R.M., *Design with Reinforced Plastics*. The Design council, 1992. ISBN 0 85072 2942.
5. Cripps, D., *Composites and Fire - a complex issue*. Ship & Boat international. Jan/Feb 1996, pp31-35.

6. Smith, C,S. and Monks, A.H., *Design of High Performance Hulls in Fibre Reinforced Plastics*. 2nd Symposium on Small Fast Warships & Security Vessels. RINA, London, May 1982.
7. Jackson, K., *Low Temperature Cure Prepregs - An Environmentally Favourable Route to Cost Effective Design and Manufacture*. Advanced Materials for Ships and Small Craft, RINA, London, November 1993.
8. Harvey, G. and Shimell, A., *Composite Construction: A Design Process Guide*. International conference on Lightweight Materials in Naval Architecture, RINA, Southampton, February 1996.
9. Rules for Classification of High Speed and Light Craft. Det norske Veritas. 1995.
10. Code of Practice for Fatigue Design and Assessment of Steel Structures. BS 7608: 1993.
11. Nicolson, I., *Small Steel Craft*. Adlard Coles Ltd, 2nd Edition, 1986. ISBN 0 229 11689 2.
12. Chalmers, D.W., *Design of Ship's Structures*. HMSO: London, 1993. ISBN 0 11 772717 2.
13. Structural Use of Aluminium. Part 1. Code of Practice for Design. BS 8118. Part 1: 1991.
14. Pollard, S.F., *Boat-building with Aluminium*. Airlife Publishing Ltd. Shrewsbury, 1993. ISBN 0 87742 377 6.
15. Rules and Regulations for the Classification of Special Service Craft. Lloyd's Register of Shipping, 1996.
16. Allday, W., *Aluminium alloys - design for fatigue*. Ship and Boat International, Sept 1993, pp 25-27.
17. The Encyclopaedia of Wood. Sterling Publishing, Co. Inc, New York, 1987. ISBN - 0-8069-6994-6.
18. Loscombe, R., *Structural Design Considerations for Laminated Wood Yachts*. RINA Conference on 'The Modern Yacht', Portsmouth 1998.
19. Structural Timber - *Determination of Characteristic Values of Mechanical Properties and Density*. BS EN 384, 1995.
20. The Gougeon Brothers on Boat Construction 4th Edn. 1985.

CHAPTER 9
STRUCTURAL DESIGN OF HULL ELEMENTS

G. Holm
Naval Architect, VTT Manufacturing Technology

9.1 THE AIM OF STRUCTURAL DESIGN

Introduction

During recent years the structural design of sailing yachts has moved into a new phase. Even though a strong empirical heritage still remains as the base for most of the series built yachts, one can see an increased use of more sophisticated methods. These are characterised by numerical strength analysis methods, a more accurate laminate properties approach and even some load determination methods with a stronger physical base.

The intent of this chapter is to describe the whole process covering direct calculation and analysis methods as well as the use and principles of different scantling determination methods.

These lecture notes are not intended to constitute a straightforward method for scantling determination. The idea is to give a guide for how to proceed with the task of dimensioning. This means that the required tools, references and the essential points which have to be taken into account are described. Established scantling determination methods offer a straightforward path.

It is expected that the reader has sufficient knowledge, or at least access to sources exposing material data and explaining the methodology of the mechanics of strength analysis.

- Define initial dimensions of structural element
- Establish load case
- Select the strength analysis method
- Define allowable deflections, stresses and/or strains
- Analyse and adjust scantlings for optimal solution

It is acknowledged that for certain projects, such as building racing yachts, where an abundance of resources are available, a much higher level of detail, complexity and accuracy can be connected to and put into the structural design process. These procedures are not always applicable in normal design work due to time and resource restrictions.

On the other hand one can without a doubt claim that many sailing disasters could have been avoided if basic common sense had been applied from the beginning, during the design process. Often simple physics is neglected in the strive for lightweight racing solutions. In fact one sometimes get an impression that people try to get around, or just 'jump over' certain physical facts.

The aim of structural design is in all cases to ensure safe operation of the yacht in the conditions and the lifetime expected.

Short overview of building materials

Wood is the oldest and the most widely known material for building yachts. From a structural point of view it has magnificent specific bending strength and stiffness in it its basic form (strength/weight). In the manufacturing process, caution has to be taken not to destroy the capacity by neglecting fibre orientation, by cutting holes in the wrong locations or by using screws and bolts of the wrong size.

In recent times more engineering-like structures have been created by using extensive adhesive bonding techniques. These allow for rational and efficient principles to be applied in both the shell structure with plywood-like solutions and in the build-up of frames and girders.

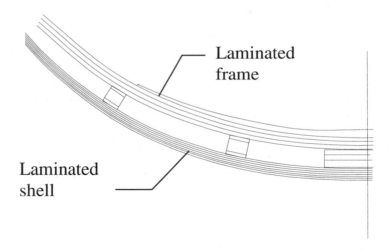

Figure 9.1 *Modern efficient wooden structure.*

FRP, or fibre reinforced plastics, is at present time the most widely used material for yachts. Its specific strength and the freedom it allows to design and manufacture stiff, efficient and repeatable structures in almost any shape, gives the material a special position in the world of yacht structures. Possibilities for the smooth alteration of fibres, including their orientation and thickness, creates unique characteristics for the structural designer to work with.

As efficient manufacturing processes like pre-preg and injection moulding techniques provide even better possibilities for high specific strength laminates, it seems that the position of FRP in yacht hulls is not threatened. For optimising purposes, however, there is a clear need to involve both skilled personnel and sophisticated software for thorough analysis.

Sandwich construction with FRP skins and foam, balsa or honeycomb cores, gives a chance to move further towards lightweight structures. If care is taken in both the design stage and during the sensitive manufacturing process, the lightest structures possible can be made using this method.

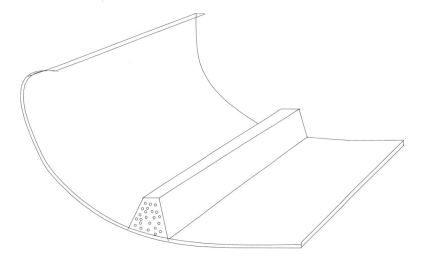

Figure 9.2 *Sandwich structures are clean simple and light.*

Aluminium is a material which plays a pronounced role in the design of bigger yachts. The manufacturing technology is more shipbuilding like, but can, in certain cases, resemble practices applied in the aircraft community. If a good finish is required there is a need for real workmanship. This is available in boatyards building luxury yachts.

From the strength point of view, problems with aluminium are almost always related to fatigue damage or galvanic corrosion. During the dimensioning process, in order to avoid fatigue failures, care has to be taken with regard to material degradation in the welding process and hard spots in the structure. On the weight scale aluminium hulls are usually at the same level as FRP single-skin hulls.

Steel hulls are easy to build, especially if developable surfaces are applied. Dimensioning is no problem either, and a lot of the bad will around steel yachts derives from the fact that they are designed by amateurs. There are a lot of good looking yachts in steel around, and their only slight disadvantage is the attention to corrosion prevention which they require. Compared to traditional wooden yachts even this fact should not be regarded as a big problem. The hull weight is about twice that of FRP single-skin hulls, so designs insensitive to weight, have to be considered.

9.2 WHAT YOU NEED

The discipline of structural design cannot be a straightforward process with clear input and readily available and usable output values. The science and nature of naval architecture with regard to sailing yachts is not in such a state. The structures are also often quite complex and a variety of load cases has to be considered. For optimal results an iterative process is often necessary. Table 9.1 shows what is needed for efficient work with yacht hulls.

Table 9.1 Needs and sources for efficient structural design work

Need	Reference
Basic book in structural mechanics including formulae for beams and laterally loaded plates	Timoshenko[1], Structural Plastics Design Manual[2], Roark[3]
Basic book in sandwich construction technology	Zenkert,[4] Allen[5]
Fundamental book in structural design of marine FRP structures	Smith[6]
Sources for load determination	Fluid dynamics principles Measurement reports with generalised conclusions, Marchaj[27, 28, 29]
References including material characteristics	Laboratory test reports; own material tests, raw material data for realistic laminates
Information on scantlings[a] in recognised reference yachts	Drawings, laminate specifications, structural details
Some good books and conference proceedings on yacht design and structures	Larsson & Eliasson[7] Delft selected Papers, SNAME, HISWA, Hammit[8]
Recognised scantling determination[b] methods	ABS[9], Lloyds[10], BV[11], GL[12], VTT-NBS extended rule[13], etc.

[a] Scantlings = yachting term for the dimensions of structural components
[b] Determination of methods (Claughton, 1998)

With respect to these sources of information, it is important to note that none of them include the whole truth. Several of them include elements which can be applied frequently. However, it seems that for a given design or yacht line, there is a need to build up one's own procedural line that includes the best suited information pieces available.

At this point it is important to note that material and especially laminate strength values offered by raw materials manufacturers are hardly ever in accordance with the laminate properties achieved in the laminate produced in ordinary boatyards. Care has also to be taken with aluminium, where the 'as welded' values have to be used in most cases.

The reason for the problems with FRP-values lies in the fact that by using ordinary production methods, the same quality of the laminate can hardly ever be repeated. In hand lay-up techniques, the excessive resin makes the difference, while in modern closed moulding methods such as Resin Transfer Moulding (RTM) and similar methods, too little resin may be applied, thus creating a laminate with inferior bending or compressive strength.

Even with basically the same type of reinforcement structure significant variations can prevail as shown in the Figure 9.3.

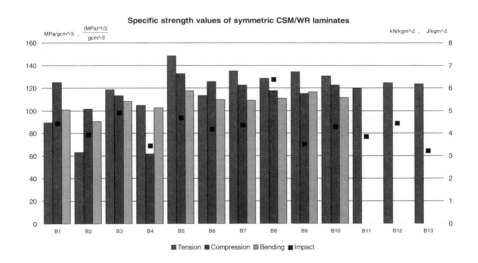

Figure 9.3 *Laminate lay-up: CSM450/WR600/CSM450/WR600/CSM450. Fibre material: e–glass. Matrix: polyester. Note the significant differences.*[14]

Many scantling determination methods have material properties listed. Some of these values may be conservative, but remember not to be too liberal before you have actual measured strength data in front of you.

168 *Sailing Yacht Design: Theory*

In conclusion, one has to have a collection of references and a procedural path to adhere to in each case. Classification rules have, however, to be adhered to from start to finish. A general sequence of actions is described next.

1. Define initial dimensions of structural element
2. Establish load case
3. Select the strength analysis method
4. Define allowable deflections, stresses and/or strains
5. Analyse and adjust scantlings for optimal solution

It is important to note that if a more detailed analysis method or other allowable strength values are introduced in a scantling determination method, the outcome will be uncertain, as the rules usually work like a package. A very useful flow chart showing different tasks when using the ABS rules is presented below.

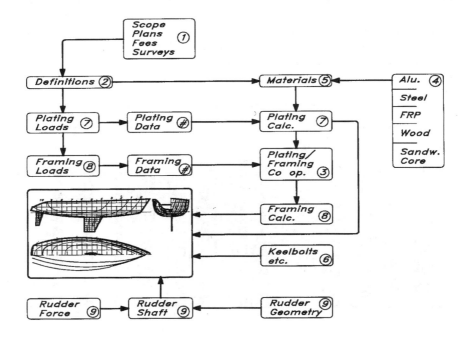

Figure 9.4 *Practical guide to the ABS/ORC rule.*[7]

9.3 BASIC STRUCTURAL DESIGN PRINCIPLES

Load, analysis method, allowables interrelationship

In the structural design process it is important to recognise the different elements included.

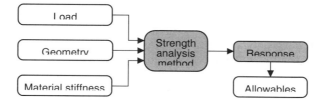

Figure 9.5 *The fundamental elements in structural design. Allowables may include stresses, strains and deflections.*

The accuracy of each one of these form the base for the accuracy of the whole process and the value of the response. If there is lack of information on loads or if the analysis method is not suitable for the case, an incorrect answer will be the result. Therefore all measures available shall be taken to ensure good quality information on all 'frontiers'.

Most of the books related to structural mechanics and the strength of materials are highly concentrated on the structural analysis side, but, alarmingly, lack information on reliable real world material characteristics, not to mention load levels on yachts.

During the recent years there has been an improvement in the availability of information. Normalised and standardised values are still in the process of being agreed on because of the comparatively young history of FRP yachts. On the metal side there is good information available about steel and aluminium, even though the fatigue values for aluminium are still under discussion.

Analysis methods

In recent years the accuracy of traditional linear structural mechanics methods have been questioned by many.[15, 16] A lot of work with efficient numerical methods has shown the significance of the non-linear characteristics of laterally loaded panels.

Linear beam and plate calculation and analysis methods are still the most widely used and they form a foundation for all dimensioning and scantling determination methods published for general use. Some methods may have some hidden coefficients or principles to take care of non-linear effects.

The principles for the use of beam and plate analogy in continuous structures are applied by analysing a single part 'detached' from the structure and assuming a representative boundary condition for the case in question. The accuracy of this boundary condition depends on the load case and the continuity of the structure. Traditionally for slow yachts static hydrostatic loads are assumed and therefore a fixed, not hinged, condition is assumed due to load symmetry over the frame or bulkhead where no angular displacement occurs.

In cases where slamming loads can occur, a much more asymmetric situation prevails. This may suggest a boundary condition somewhere between fixed and simply supported. This is often used, especially for deflection analysis. For the stress analysis a conservative condition may be used.

As earlier stated, significant non-linearity exists in the response of not only laterally loaded single skin panels but also in sandwich plates. This means that deflections and stresses do not increase in direct proportion to the load. The effect is due to membrane effects at higher deflections values, where some of the bending load is carried by tensile stresses as if the plate or panel is working like a tensile canvas. This effect is also present in simply supported plates without a stiff frame. If the plate, however, has stiff boundary conditions along the edges, the effect is more pronounced.

The utilisation of this effect is not easy. In work performed by VTT it has been shown that weight reductions in the order of 40% can be achieved by optimising sandwich structures with methods accepting non-linear response analysis. In these cases all stress components maintain sufficient safety factors while no limits are put on maximum allowable deflections.

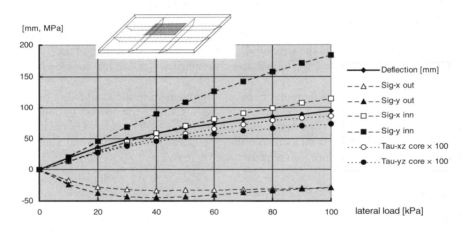

Figure 9.6 *Deflection and most critical stresses of the mid-panel. Lateral load on the mid-panel. Note that the deflection is the difference between the highest (at the middle of the mid-panel) and the lowest (at the corners of the mid-panel) deflection value of the mid-panel.*[15]

If inplane loads are introduced simultaneously with the lateral loads a different situation is created. If large deflections are allowed these structures have to be treated with utmost care, as this may lead to increased stresses in them during the prebuckling stage.

In literature one can find iterative procedures by which membrane effects can be taken into account in plates. The structural design procedure is thus slowed down. For thorough analysis it is advised to use a numerical finite element software capable of non linear calculations.

Dimensioning methods

Dimensioning methods for yachts have been published for a long time. The most famous are perhaps the traditional Herreshoff and Nevins rules for wooden yachts. These were clearly explained in the excellent classical book of Skene/F.S. Kinney 'Skenes Elements of Yacht Design'.[19] Later Gibbs and Cox published rules and guidance for the dimensioning of FRP–yachts. More strict and systematic rules have been published by many classification societies later on, and rules for many different materials can be found in their rules.

Recently the ABS rules have been widely used in the offshore racing community. The popularity was created through co-operative measures by ORC and ABS where these rules were considered mandatory for some racing yacht categories. At this stage ABS has disregarded yachts under 24 m and there is work underway to create an international ISO standard for yacht scantlings. The first public drafts of this standard will be available during 1998.

Other rules are published by all classification societies such as Lloyds Register, Bureau Veritas, Germanishe Lloyd, RINA and DNV/NBS. VTT in Finland has extended the applicability of the Nordic NBS rules to 24 m and has also introduced the design categorisation required by the Recreational Craft Directive (RCD).

An important point with all these rules is that they are **dimensioning** rules. This means that they are not designed nor intended for strength analysis. This means that in most of them the load, the strength analysis method, and the allowables are cooked into one streamlined package. The result and the outcome of a calculation is a measure or a dimension requirement of a frame. The formulas cannot and should not be used separately. The loads are usually considered as **design** loads and are sometimes only loosely related to actual loads.

The analysis principle and the allowables are also integrated, so that no conclusions can be made on behalf of real values or assumptions of these. For a person more actively interested in these kinds of strength analysis procedures, this has been a sad development. Fortunately, a more open structure is designed into the VTT NBS Extended rule and the new ISO standard. The ABS rule was the first significant step towards this more engineer-friendly development, which facilitates direct improvements along with improved knowledge in the different areas.

These rules can be applied and are especially applied when the customer requires scantling determination or complete certification of the yacht in relation to a certain set of rules. You need practice to be able to work with them with confidence. This is because definitions and procedures may vary between them. This means that, sometimes, when using one principle in one rule you have to use another in the other one.

Examples of differences are: how to measure the frame spacing, the factors of safety, and the calculation of effective plate flange for frames and so on. A brief description of the different rules is included in the following table.

Table 9.2 Rules and their characteristics

172 *Sailing Yacht Design: Theory*

Rule	Notes
ABS/ORC[9]	Engineering approach, stiffness usually governing, high allowable strength values (never 'pushed' ?)
Lloyds Register[10]	Rule frame spacing oriented, no allowables, new rules for craft > 24 m
Germanishe Lloyd[12]	Detailed, sandwich, cold moulded wood and rig dimensioning included
Bureau Veritas[11]	Focus on bending moment resistance of laminate, rigs included
RINA[17]	Clear concept, not detailed
DNV / NBS	Practical, works for small yachts, membrane effects included empirically
VTT NBS Extended[13]	Clear structure, real values and principles, quasi membrane method, recreational craft directive adapted design categorisation.

Basic principles of dimensioning methods

Almost all rules are built around three formulas derived from linear beam and plate theory. A direct procedure by a structural designer should also cover at least these, too. They are minimum thickness, strength of plate and stiffness of plate. The formula are of the format:

$$t_{min} = k_1 L + k_2 \quad \text{minimum thickness} \quad [9.1]$$

L = length of yacht
k_1 = coefficient
k_2 = basic minimum value of thickness

Other parameters may be introduced too, such as displacement and strength of material.

$$t_{strength} = k_1 s \sqrt{\frac{k_2 p}{\sigma}} \quad \text{strength} \quad [9.2]$$

s = frame spacing
p = pressure load
k_1 = a factor for curved panels and boundary conditions
k_2 = a coefficient for aspect ratio and the material safety factor

$$t_{stiffness} = k_1 s \sqrt[3]{\frac{k_2 p}{k_3 E}} \quad \text{stiffness} \quad [9.3]$$

k_1 = a factor for curved panel, boundary conditions and a deflection limit
k_2 = a coefficient for aspect ratio

k_3 = a factor for load distribution
E = stiffness of material

For sandwich panels and frames there are requirements on section modulus SM and moment of inertia I. For FRP laminates the SM is calculated with reduced values keeping a standard reinforcement type as a basis. The rules usually handle laminates in such a macroscopic way. By using a more thorough FRP analysis procedure each lamina may be assessed separately.

A complete guide for the use of the ABS/ORC rules is included in Chapter 12 in Larsson & Elisasson.[7]

9.4 LOADS

Pressure loads

Hydrostatic loads

Hydrostatic pressure loads have historically been the basis for all dimensioning procedures because dynamic effects only come with speed and high performance. Usually, slow craft decks, and bulkheads are dimensioned against a static pressure height. Also, in many cases dynamic loads are reduced to representative static loads for simplicity.

A peculiarity under the hydrostatic headline are the transverse rigging loads that are induced by the stability of the yacht and of the pretensioning of the shrouds. Without stability no shroud loads!

Hydrodynamic loads

A typical example of the above mentioned is the hydrodynamic movement of a yacht in waves, with the hull in and out of crests and troughs. This creates a changing so called hogging and sagging bending moment on the hull. This is small in a yacht, especially as these have very slender bows and sterns in comparison to the midship part.

Other hydrodynamic loads are created by the keel and rudder, on which hydrodynamic lift is created. This lift causes bending moments in the rudder stock and in the keel attachment. By establishing flow speeds and angles of attack these forces are definable through fluid dynamics principles.

Slamming loads

As the yacht is moving in a rough sea the bow occasionally falls into a wave with a loud bang. In these situations a slamming pressure usually sweeps over the bottom and side shell, creating a rapid pressure peak with a falling pressure afterwards. Very little has been published on actual measurements so far. Both Hentinen & Holm[18] and Larsson & Eliasson show values that are very high.

174 *Sailing Yacht Design: Theory*

The measurements on board s/y Sail Lab indicate pressures in the order of 50% more than the ABS/ORC design loads. These rules have since then been amended to reflect extremely high pressures on offshore racing yachts.

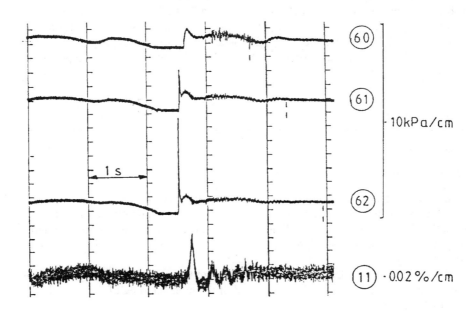

Figure 9.7 *Response of pressure gauges and in the stringer about 1–1.5 m in front of the mast on board the 9.5 m laboratory sailing half tonner Sail Lab.*[18]

In the bow area in front of the mast it is advised to adopt a safe approach to the structural design and utmost care has to be taken.

Because the uneven distribution of the slamming load over the surface, a quasi static 'even distribution' approach is usually applied. This means that the smaller the structural element in question is, the higher is the mean pressure expected to act on it. Here we come to the so called pressure reduction curve, which relates the maximum evenly distributed pressure to the design or load area in question. Several methods have this included and presently the method expressed by Allen–Jones[31] is favoured by the ISO working group.

The curve expressed in absolute terms on the basis of measurements on board s/y Sail Lab is displayed in Figure 9.8 together with the curve derived from ABS/ORC. The shape is the same but there is a difference in level.

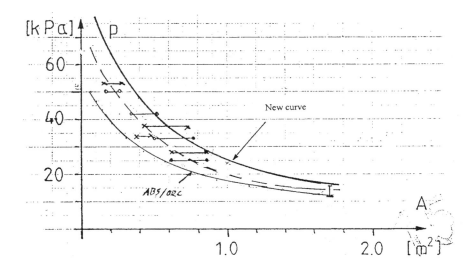

Figure 9.8 *Pressure reduction curves for a 9.5 yacht. Values are according to measurements and calculations.*

It is not possible to give a complete picture of all the different loads, which have to be applied to yachts. Therefore an example is given how the load is established for one detail in one of the scantling determination rules. According to ABS/ORC the design head for the side shell in the bow area is calculated as:

Head = 1.08(hd – d) m

hd = 3d + 0.14L + 1.62 m

    hd = basic design head
    d = canoe body depth
    L = rule length of yacht

The pressure head is then reduced with respect to the area reduction factor. The so called pressure head is expressed in meters of water hydrostatic pressure, a fact that has historical background. It gives a value 9.81 times or approximately ten times greater in real pressure units (kPa).

In the NBS VTT extended rule, the base pressure is given in a graph as a function of the speed and length of the hull. This is then corrected for actual displacement, longitudinal and transversal location and the design category.

Global loads

Hull girder

It is usually not necessary to consider global hull loads for ordinary cruising yachts and yachts built in single skin laminates. The layout of these yachts gives a fairly high hull beam with ample geometry to give a high moment of inertia and stress well under allowable strength values. A low neutral axis from thick bottom shell laminates and longitudinal girders is not enough to create high compressive strains in the deck structures.

However, there are several cases where the structure has to be checked against global bending loads. In these cases, the slamming and/or rig induced longitudinal loads induce stresses, mostly in the deck area, which can cause cracking in critical areas of the deck. One possible maximum bending moment situation is of course the maximum breaking strength of the fore- and aft-stays. It is reasonable to expect that these would break before the hull breaks. For transient slamming load situations normal naval architectural principles for the calculation of the bending moments can be used on a sailing yacht too. This is hardly ever done, however.

Attention has to be paid in the following cases:

- long slender high-speed hulls (above 15–20 m)
- low profile hulls
- hulls with very thin faced sandwich structures
- decks with large openings such as doghouse type superstructures
- decks where series of hatches are located transversely

For the hull to be able to carry global loads it is important that longitudinal continuity is ensured. This means that laminates, stringers and deck shapes must have a continuous structural 'flow' in order to ensure the efficient transference of stresses. In any location where these have discontinuities, there is a risk of cracking or failure.

Typical cracking areas are hatch opening corners, window areas, stringer and deck beam connections. These are consequently the areas which have to be reinforced to eliminate stress concentrations.

In addition to strength considerations, it is important to note that racing yachts especially, shall have a hull with ample stiffness to be able to carry the rig efficiently. The rig, a very stiffness sensitive structure, is dependent on a rigid foundation to work properly in close-hauled sailing conditions.

Clean continuous shapes are stiff. For the box girder type beam which the hull constitutes, it is important that the transverse sections do not distort, thus creating a flexible beam. From time to time internal backbones are built throughout the hull or in the bow or stern part of the hull only. These spaceframes are often designed in such a way that they efficiently carry the local rig chainplate loads in addition to mast foot and keel loads. The hull 'keeping the water out' is attached to the frame.

Except for extreme load carrying tasks around shrouds and mast foot, it is usually more weight efficient to design an integrated hull structure without the spaceframe. Such an integrated structure comprises stiff shell laminates and closely spaced longitudinal stringers on web frames and light bulkheads. These 'integrated' structures are familiar in for example aircraft, where the spaceframes were abolished long ago.

Local loads

Keel

Traditional long keels do not pose a structural problem. Things are different with deep fin keels not to mention the extremely short, deep and thin fins with a large amount of weight in a bulb deep down. These create a large moment at the root of the cantilever where the keel is fastened.

The static transverse bending moment created when the yacht heels to 90° is often used as a basic load case. This may be increased by dynamic effects. In normal sailing conditions the weight is counterbalanced by the hydrodynamic lift when underway.

In addition to this, a reasonable guarantee against grounding loads has to be included in the strength assessment. A design speed or the characteristics of a suitable standard grounding case is still to be decided on by the structural community, but a full stop situation in about 3–4 knots without damage could be a base for discussion. During grounding tests performed with the laboratory yacht 'Sail Lab', an angle of around 15° of solid granite rock was surprisingly experienced as full stop.

A criteria like the one proposed could then ensure significantly higher speeds when hitting small stones, sand or mud. ABS/ORC have a load for these situations that is:

- Longitudinal load at forward bottom edge of keel and interpolated:
  Lwl > 20 m; 3 × Displacement (force)
  Lwl < 10 m; 1.5 × Displacement (force)
- Upward force = 1.5 × Displacement (force)

The structure to which the keel is attached have to be dimensioned against these loads.

Rudder

The structure around the rudder stock is more easily dimensioned when the bending strength of the rudder stock can be used as the required load. To be able to carry this bending strength, the bearings have to be located sufficiently far apart. In addition the structure 'between' the bearings has to be stiff enough to carry the transverse load components on the bearings.

As long as the speed and the lift coefficient are agreed upon, the dimensioning of rudder stocks is a straightforward procedure. We have clear indications that as the speed increases, the maximum attainable lift coefficient decreases. This may be due to limitations in resisting torque by helmsmen or due to the turning response of the yacht. The phenomena has been documented.[18]

178 *Sailing Yacht Design: Theory*

The levels proposed in many methods seem adequate.

Impact

Both the deck and hull may be exposed to local impact loads. Falling spinnaker booms or winch handles hitting a window or, alternatively, a floating object hitting the hull at high speed, create a demand for sufficient local strength. Traditionally this has been met by applying different minimum thickness requirements on aluminium, FRP single-skin laminates or sandwich outer face laminates.

The actual value of the minimum thickness varies between rules and the location of the structural element in question. Recent work done by VTT and others has shown that by linking this requirement to reinforcement weight, a more accurate picture of the real impact strength of the laminate is provided. This is the principle in the ABS rules.

Unfortunately no agreed value of a typical impact energy has been set. This would help a lot, allowing specific typical thickness values to be drawn. Comparisons between single skin and sandwich structures is also difficult. While there is a need to limit surface damage to a minimum on cruising yachts, criteria for complete penetration energies may be the only applicable criteria for pure racing yachts. In some cases it seems that the impact resistance criteria does not exist at all for these yachts.

9.5 PRINCIPLES OF EFFICIENT STRUCTURES

Basic principles

To be able to design a weight optimised structure one has to follow certain 'weight efficient' principles. Closely following these principles will minimise the risk for overdimensioning and will keep the strength level the same in all details.

Structural Design of Hull Elements 179

Table 9.3 Optimal solutions in an efficient structure

Detail	Demand	Optimal solution
Materials selection	Strength/Weight	Uni-directional (UD) and/or stitched reinforcements, Carbon if lightness crucial otherwise glass, Epoxy or Vinyl ester used in the appropriate way, material compatibility, fibre content
	Stiffness	Optimised direction of fibres, high modulus fibres
	Impact	Glass, strong core, flexible matrix, elastomer coating
	Durability	High quality resin, surface treatment, marine grade aluminium, corrosion resistant steel
Panel laminates	Strength/Weight	Small thin single-skin panels, optimised fibre direction, high strength fibres, epoxy
		Sandwich is light but locally weak
Frame spacing	Optimum	Small to a limit – optimisation possible
Frames	Low weight	Longitudinal, few and high, +/–45 in webs, UD in flanges, efficient secondary bonding
Bulkheads	Arrangement	Light sandwich holes – but not impairing strength, weight efficient connection to hull and frames
Keel attachments	Strong and light	Floors continuous, tapered and attached to stringers, no hard spots
Interior	Lightness	Integrated – not doubling strength elements, if not load carrying – then completely loose
Deck layout	Continuity	Continuous straight shapes, no 'spring effect' shapes, local strong points for point loads
Dimensions	Low weight	Low density and thick, is strong and stiff in bending, unidirectional properties in flanges, shear stiff in webs

Impact resistance

The strength of laminates against impact loads can be increased by several different means. Figure 9.9 shows some influences of material selection on the specific impact strength.

180 *Sailing Yacht Design: Theory*

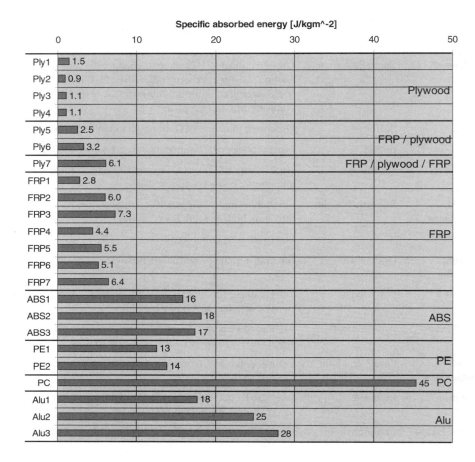

Figure 9.9 *Comparison of specific impact strength (absorbed energy per square weight) of tested materials.*[20]

Emphasising the differences between FRP laminates, Figure 9.10 shows the difference between glass and aramid fibres.

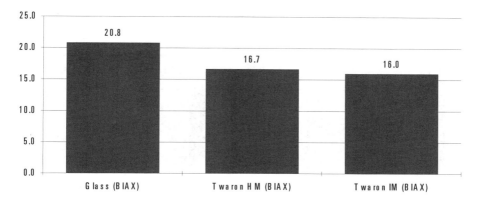

Figure 9.10 *The effect of the fibre material on the specific impact strength [J/(kg/m^2)] of sandwich panels. Face laminate lay-up: CSM450/BIAX600(E–glass)/CSM450, CSM450/BIAX(Twaron HM)220/CSM450, CSM450/BIAX(Twaron IM)350/CSM300. Core 20 mm Airex R63.80.*[21]

Glass fibre laminates perform quite well in comparison to aramid and carbon fibre laminates on an equal weight basis. Special hybrid laminates have a very low impact resistance. The fact that aramid fibres have a good reputation may lie in the fact that this is true for high speed ballistic purposes where special aramides like Kevlar 29 are used. In these cases fibre surface sizing gives a loose fibre to resin connection and a lot of energy can be released. In boat hulls, however, stiffness is needed and other material systems are needed.

Bad impact strength values can be found especially in hybrid aramid/glass or carbon/glass laminates. The fibres seem to act in a very unfavourable way in these cases. Several reports on these effects have been published.[22,23]

Finally it must be remembered that the impact strength of a sandwich panel is strongly dependent on the core material. Figure 9.11 gives an impression of this effect.

182 *Sailing Yacht Design: Theory*

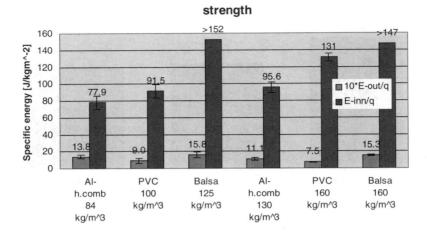

Figure 9.11 *Effect of the core material and core nominal density on the specific impact strength of sandwich panels. Note that, for clarity, the values at penetration of the outer face are scaled by a factor of ten.*[22]

9.6 SAFETY FACTORS

Different methods and their principles

Safety factors constitute a considerable area of different approaches. The idea is to cover different uncertainties in the different areas of the procedures. The optimal situation is when all uncertainties are known and documented. In such a case a precise FoS can be calculated. Unfortunately the scantling determination world is filled with uncertainties. A picture of the situation in the FRP field can be gathered by looking at the 'cosmic perspective' offered by Figure 9.12.[24,25]

Structural Design of Hull Elements 183

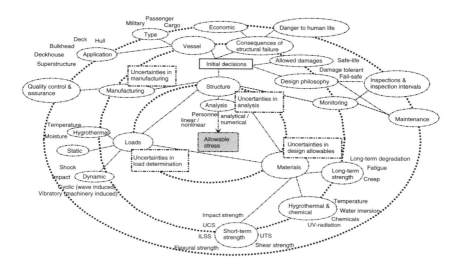

Figure 9.12 *Areas where uncertainties may be created or just happen.*

Generally there are three methods for establishing the allowable stresses:

- A reliability-based approach.
  □ The method requires statistical data of loads and of structural response including material behaviour.
- Applying a single Factor of Safety (FoS).
  □ It is the most common method. FoS are derived from experience and the value is usually a 'rough' estimate.
- The third method is to apply partial factors of safety.
  □ PFS´s method is an enlargement of the single FoS method.

Usually there is a considerable lack of information for the reliability and the partial FoS-methods. Therefore, a single value is usually selected covering all aspects of uncertainties. In the case of scantling determination methods, additional FoS may be applied by selecting, for example, conservative coefficients in the formulae, but showing one in relation to the material strength values.

The possibility to use the partial safety factor method is attractive as it facilitates a chance to evaluate all the different aspects separately. This makes the whole problem more transparent and provides a better chance to focus on the real problem areas.

As partial factors traditionally are multiplied in the PFS method a very high value is obtained. This is because 100% safety is achieved when the worst case is assumed for each factor. The factors are in addition assumed to be independent.

On the basis of the above-mentioned, VTT has developed a method and a PC-program to calculate a FoS assuming that the probability that all factors achieve their maximums simultaneously is very low. The method seems to give reasonable values in addition to showing where additional efforts have to be put to increase certainty.

With the question of more accurate FoS evaluation rises the safety philosophy. Is a safe-life, a fail-safe or damage tolerant approach to be applied? In addition, the consequences of a failure have to be assessed. Human life, large or small economical consequences, service intervals, efforts and costs are the issues to judge on.

Typical values of single FoS used in some references

The following table contains some information on FoS included in different rules.

Table 9.4 Typical single factors of safety

Rule	Element	In relation to	Values
DNVLC [26]	Bottom panel in slam. area	First Ply Fail (FPF)	1.5
		Last Ply Fail (LPF)	3.3
		σ_{nu}	3.3
	Sandwich core	τ_{ult}	2.9
	Other structures	σ_{nu}	3.3
		τ_{ult}	2.5
	Long term static load	σ_{nu}	5.0
		τ_{ult}	6.7
ABS/ORC	Shell and deck plating	St and Al ult. tens	1.7
		Al as welded	
		FRP ult flex	2.0
	Beams and frames	St and Al ult. tens	2.0
		FRP tens flange	2.0
		Compr. plating	2.0

In cases where the laminates have transverse fibres included first ply failures occur usually at very low levels. The strains at first ply failures and microcracking are typically between 0.2–0.5% for CSM and mat/woven rowing laminates. This means that the ratio of ultimate strength to first ply failure is between 2.7 and 5.3. This argues for high safety values against ultimate strength. If high quality resins are used it seems that the strain at first ply failure is increasing slowly.

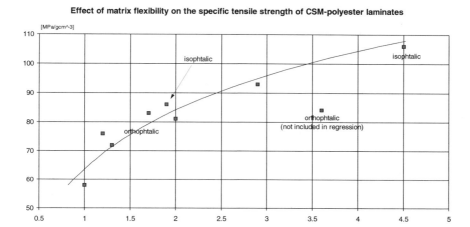

Figure 9.13 *Matrix flexibility and specific tensile strength for ordinary low modulus boat building chopped strand laminates.*[14]

9.7 DEFLECTION CRITERIA

Deflections can be governing

Deflection limits were introduced in the scantling determination process when FRP structures started to be applied. This was because of the relatively low stiffness of these materials. Especially decks on which people walk are sensitive areas, and traditionally a deflection limit of 0.5% of the span has been applied when walking loads are applied. It is claimed that this satisfies the need for a stiff deck.

Deflection limits of 1–1.3% are traditionally applied to other hull structures. There is an ongoing discussion, whether this is relevant or not. Many think that if there is no specific functional need to limit the deflection, no requirement or a much more liberal requirement is acceptable as long as the strength is sufficient.

Deflection limits tend to keep the frame spacing short, thus also keeping the structure closer to linear behaviour and in many cases also the stresses quite low. If the strength related FoS in these cases are high it is a waste of material and the structure will have excessive weight.

The best solution is to carefully examine if a criteria is necessary or not. If applied, excessive values have to be avoided if they are not required. If large

9.8 HULL SHELL DIMENSIONING

Lateral and in plane loads

Hull and deck outside shell panels are dimensioned using traditional plate theories. Evenly distributed load cases are listed in almost all references but local load cases are rarely defined. The most critical stresses have to be identified in all relevant directions and compared to allowable stresses. In addition, the deflection values have to be checked if appropriate. For aluminium there are methods by which a certain permanent set can be taken into account if this can be allowed.

In single-skin FRP structures the real bending stiffness and section modulus of the plates is the most difficult information to quantify. The stacking sequence of the laminate and the actual fibre reinforcement percentage are the key factors. There are several references explaining the principles of laminate theory,[30] and some programs exist for calculating laminate stiffness'. Some typical values for common laminate types are often shown graphically as a function of fibre content and/or fibre main directions.

In sandwich structures the shear loads are carried almost completely by the cores. Therefore the shear stresses have to be checked, as do the additional deflections created by shear deformation of the cores. Very thin face sandwich strength analysis include in addition the control of several different failure modes of the face structure.

In certain cases simultaneously acting in-plane loads have to be included in the analysis for accurate results.

Strength with respect to local loads

The capacity of a certain structure to carry local loads is a more difficult calculation. Firstly, as already stated, the strength criterion is not clear. Secondly, very few theoretical methods are available to calculate the strength value of plates.

The most accurate way is through tests to check whether a new suggestion produces the same level as one which empirically has given confidence. Actual boats or the minimum thickness values in the scantling determination methods may serve as bases for these kind of evaluations together with a representative test. A working test method with a pyramid shaped impactor has been introduced by VTT.[22]

Non-linear effects of plates and plate fields

The non-linear effects described previously introduce new problems in the field of dimensioning of yacht structures. At present there is no established way of calculating, except for numerical finite element analysis. Riber[16] has developed a closed form analytical solution by means of energy principles. It seems to be a promising way of dealing with this phenomena and can possible contribute to a procedure for easy non-linear dimensioning.

The author is convinced that inaccuracies in linear methods are of such a magnitude that a major step towards non-linear methods has to be taken.

In highly loaded sandwich plates non-linear analysis show that the shear stresses are not reduced by the same amount as the normal stresses. This means that when these panels are overloaded and large deflections occur there will be core failures before there are face failures if the sandwich panel is designed with linear theory.[15]

This is just in line with practical experiences in yachts, even though manufacturing problems often are inferred when 'delaminations' and core failures occur.

Complex structures

Many yachts have peculiar shapes, structural styling and design features that are impossible to simplify accurately to make simple straightforward strength analysis easier. In cases where shell-type and open profile-like structures have to be designed empirical judgement, conservative values or sophisticated numerical analysis are the only solution. Full scale tests then have a much more critical role than in the case of more ordinary structures.

9.9 DIMENSIONING OF FRAMES AND FRAME SYSTEMS

Load distribution

Longitudinal frames are primarily dimensioned as single structures with a length between web frames or bulkheads and loaded half way to the adjacent frames on both sides. Appropriate area reduction has to be applied. Both normal stresses and shear stresses in the webs have to be included in the procedure.

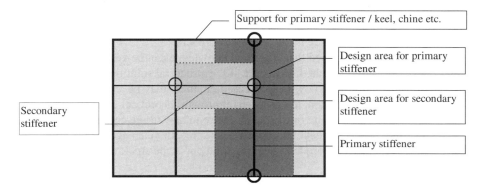

Figure 9.14 *Principles for design load area definition.*

Effective flange and frame construction

For analysis of stiffener and grid systems the reader is referred to excellent descriptions by Smith.[6] The book includes also the important elements of shear and buckling considerations needed in optimised structures. In ordinary cruising yacht structures buckling is seldom a problem but with high stiffeners with strong laminates in the flanges and thin webs this may be a major consideration.

Connections and joints

Connecting structural members to the shell and to each other requires detailed analysis. The scantling determination method usually includes a lot of good advice on how this should be done in the manufacturing process, well aware of the fact that these are seldom analysed. Therefore solutions are offered that include adequate margins of safety. If, however, an optimised solution is required, mostly shear stress analyses have to be performed.

A lot of research has been performed on good principles for the attachment of frames through different so called secondary bonding techniques. The results of these investigations can be applied in many cases. The number of joints in a yacht is considerable and there is a lot of weight and work hour savings possible.

In areas with two frames or stringers crossing, the continuity of sufficient low shear stresses has to be ensured by doubling web laminates or by continuity of the stringer through the higher frame.

9.10 CONCLUSIONS

It has not been considered appropriate to include the formulae required for strength analysis in this chapter. These are nowadays available quite easily in the literature The focus has been on what considerations have to be made during the analysis. As can be seen there are both problems and possibilities available.

As in many other areas attention to detail is the key issue here, too. If this attention is too focused on one or a few details and the overall touch to the strength problem is lost, there will almost certainly be problems. A broad and evenly distributed degree of accuracy is the optimal approach.

Finally it has to be stressed that empirical knowledge can be very useful and should never be underestimated. However, if the structural solution deviates from the solutions the empirical knowledge is based on, there may be a problem. Therefore care has to be taken, to ensure, by thorough theoretical analysis, the reliability of the new structure or the old structure in new usage conditions.

The disasters with the breaking rudder stocks some 15 years ago, proved this. The change of focus from torque to bending moment in the stock was not recognised clearly enough.

REFERENCES

1. Timoshenko, S. and Woinowsky-Krieger, S., *Theory of Plates and Shells.* McGraw-Hill Kogakusha, LTD.
2. *Structural Plastics Design Manual.* Phases 2 and 3; Chapters 5 - 10, U.S. Department of Transportation, Federal Highway Administration, Urban Mass Transportation Administration. 1982.
3. Roark, R.J. and Young W.C., *Formulas for Stress and Strain.* 5th ed. McGraw-Hill, 624, 1975.
4. Zenkert, D., *The Handbook of Sandwich Construction.* North European Engineering and Science Conference Series. , 442, 1997.
5. Allen, H.C., *Analysis and Design of Structural Sandwich Panels.* Pergamon Press, 283, 1969.
6. Smith, C., *Design of Marine Structures in Composite Materials.* Admiralty Research Establishment, Dunfermline, Fife, Scotland. 389, 1990.
7. Larsson, L. and Eliasson, R., *Principles of Yacht Design*, Adlard Coles Nautical, 302, 1994.
8. Hammitt, G., *A. Technical Yacht Design.*, Adlard Coles Nautical, 238, 1975
9. *Guide for Building and Classing Offshore Racing Yachts.*, American Bureau of Shipping & Affiliated Companies. 1994.
10. Leadbetter, W., *Rules and Regulations for the Classification of Yachts and Small Craft..* Lloyd´s Register of Shipping, Yacht and Small Craft Department. Notice No. 1. 10.6.1982.
11. *Rules and Regulations for the Classification - Certification of Yachts.* Bureau Veritas, International Register for the classification of ships and aircraft. March 1990.
12. *Klassifikations- und Bauvorschriften Schiffstechnik.* Germanischer Lloyd. TEIL 3 - WASSERSPORTFAHRZEUGE Kapitel 1 - 7 Ausgabe 1991.
13. Furustam, K-J., NBS-VTT Extended Rule,. VTT Technical report VALB172, 17 pp 16.12.1996.
14. Hildebrand, M. and Holm, G. *Stronger laminates in production boats. Material parameters.* October 1991. VTT Research Notes 91/1289. 54 p. + app. 42 p.
15. Hildebrand, M. and Visuri, M. *The non-linear behaviour of stiffened FRP-sandwich structures for marine applications.* 21.8.1996. VTT Technical report VALB155, 53 pp.
16. Riber, J. H. *Response Analysis of Dynamically Loaded Composite Panels.* Department of Nava Architecture and Offshore Engineering Technical University of Denmark. 172 pp., 1997.
17. *Rules for Pleasure Vessels.* Registro Italiano Navale, 1973.
18. Gerritsma, J., Heer de P.W., Keuning, J.A. Selected Papers HISWA Symposium on Yacht Architecture 1969–94. Delft University of Technology Ship Hydromechanics Laboratory Delft, The Netherlands. 626 pp.
19. Skene/Kinney, S.F., *Skene´s Elements of Yacht Design.* 351 pp.
20. Hildebrand, M. *Local impact strength of various boat-building materials..* VTT Publications 317, 28 p. + app. 6 p., 1997.
21. Hildebrand, M. *The effect of raw-material related parameters on the impact strength of sandwich boat-laminates.* VTT Publications 211. 36 p + app. 19 p. 1994.
22. Hildebrand, M. *Improving the impact strength of FRP-sandwich panels for ship applications.* VTT Technical report VALB138, 42 pp. 1996.
23. Aamlid, O., Antonsen. A., *Oblique Impact Testing of Single Skin, Aramid Fibre Reinforced Plastic Panels.* Det Norske Veritas. 30.January, 1997.

24. Visuri, M. *Allowable Stresses in FRP Marine Vessels*, 14.1.1995. M.Sc. Master's Thesis, 100 pp.
25. Visuri, M. Corfos -*Program 1.0. factors of safety determination tool for composite ship structures*, 9.1996. VTT, Technical Report VALB156, 51 pp.
26. Det norske Veritas. *Tentative rules for classification of high speed and light craft*. 1991.
27. Verdier, G. *Method of Analysis and Design of Sailing Yachts Structures*, DTU, 140 pp. 1996.
28. Holm, G. *En planande båts bottenkonstruktion. Dimensionering och materialjämförelse*. Januari, 1980, 101 pp.
29. Davies, P. and Lemoine, L. *Nautical Construction with composite materials*, 7.-9.12.1992 International Conference Paris, 467 pp.
30. Jones, Robert M. *Mechanics of Composite Materials*. New York:Hemisphere Publishing Corporation. 355 p. 1975.
31. Allen, R.C & Jones, R.R., A Simplified Method for determining Structural design limit pressures on high performance marine vehicles. AIAA Conference, 1978.

CHAPTER 10
MAST AND RIGGING DESIGN

D. Boote
Dipartimento di Ingeneria Navale e Tecnologie Marine,Università di Genova

10.1 INTRODUCTION

While in the past the design of sailboats was undertaken on the basis of semi-empirical methods and of the personal experience of the naval architect gained on previous vessels, nowadays the introduction of powerful design tools and methodologies, mainly based on computers, has led to very high performance sailing yachts. Especially for racing yachts big efforts have been made by many research centres and technical offices; it can be said that the success and the value of a sailing yacht is tightly bound to its racing victories and this is enough to justify the effort to improve design methodologies. Together with the progress in hydrodynamics and aerodynamics, yacht structures also had to follow this trend. The employment of new materials and the utilisation of sophisticated methodologies for structural analysis allow improved strength and safety characteristics, reducing, at the same time, the weight of the vessels. As part of yachts structures mast and rigging should be considered with particular attention because of the influence they have on the yacht performance. Granting that a mast should have the adequate resistance to withstand the most (reasonably) severe loads transmitted by sails, a well-designed mast will result in:

- Reduced spar section with better windage characteristics of mast and rigging
- Lighter weight and lower centre of gravity of the boat
- Reduced roll and pitch movement of the boat with a considerable increase in boat performance and comfort

The aim of this chapter is to assess the problems inherent to the structural design of mast and rigging and to present synthetically the methods actually employed for their scantling determination.

10.2 MAST AND RIGGING

If the sails system has the fundamental function of transforming the wind velocity into propulsive force for the vessel, spars and rigs have the equally important task of controlling the optimum sail shape. The mast, in particular, can be considered the most important element which bears the higher loads transmitted by sails; the boom mainly controls the attack angle of the mainsail and is subjected to lower loads. Sails are able to transmit very high forces to the mast and, when it is unsupported, it can experience very high displacements at the top. This would have the consequence of lowering the pressure distribution on the sails and decreasing the propulsion of the boat. In order to resist these loads the mast should have very large structural section, which would involve poor windage, heavy spar weight and high centre of gravity.

Even if many kinds of small sailing boats are fitted with unstayed masts, for other sailing yachts the mast should be supported by a three-dimensional rigging system fitted athwartship (shrouds) and in the fore and aft plane (stays). The rigging is attached to the mast at different levels and secured to the yacht hull; the shrouds are secured to the hull by chainplates and the stays are connected to reinforced hull points at the bow (headstay) and the stern (backstay). The constraining action of rigging depends on the angle β the wire forms with the mast; the tension T in the wire acts like a constraint for the mast for an amount $T \times \sin\beta$ and stresses the mast with a compression equal to $T \times \cos\beta$ (see Figure 10.1(left)). The wider the angle, the lower the compression on mast. For this reason angles below 10 or 12° are not recommended.

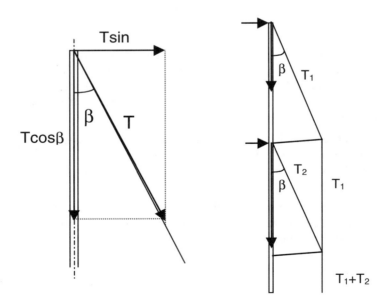

Figure 10.1 *Mast supports: longitudinal (left); transverse (right).*

While this is not a problem in the longitudinal plane because of the space availability, which allows opening up the stay angle 30° and more, a different situation exists in the transverse plane where the maximum shroud angle is limited by the breadth of the hull. For small boats with short masts a single shroud can be fitted with a minimum aperture of 10°. In the case of higher masts it is no longer possible to respect this requirement and it becomes necessary to open the shrouds with quasi-horizontal rods called spreaders whose maximum length depends on the hull breadth. More than one set of spreaders can be fitted as a function of mast height in order to maintain the angle β above 10°. At each spreader order shrouds give support to the mast but, at the same time, compression on mast increases; Figure 10.1 (right). The shrouds can be continuous or discontinuous; the continuous solution consists of full-length shrouds from the mast attachment point down to the chainplates. In this case the wire section is constant for the whole length. The discontinuous solution consists in separate spans connected at the spreader end with mechanical links.

Figure 10.2 *Masthead and fractional rig arrangement.*

In the longitudinal plane the need to trim the mainsail by boom rotations requires a large open space so that it becomes difficult to set support points for the mast at intermediate heights. The way the mast is supported depends on the type of rig: in a masthead sloop the mast is sustained by a headstay fore and a backstay, while in a fractional sloop the mast is sustained by a forestay, by a backstay attached to the top of the mast and by running backstays attached to counteract the forestay. In the cutter configuration, typical of bigger yachts, the mast has an additional support ahead, a babystay, and, optionally running backstays after (see Figure 10.2). The presence or not of running backstays depends on the nature of the yacht: in a cruising yacht it is preferable to avoid the runners in order to make the boat easier to handle, while it is necessary to set them on a racing yacht in order to better trim the mast and achieve best performance. For all the configurations

194 *Sailing Yacht Design: Theory*

considered spreaders can be set in line with the mast axis or swept aft in order to give additional support in the longitudinal plane.

The type of arrangement heavily influences the performance of the boat and the strength of the mast as well. So it is very important to consider adequately the proper configuration of spars and rigging to ensure good performance.

10.3 STRUCTURAL ASPECTS

Mast, boom, stays and shrouds form what is known as 'standing rigging', that is the category of equipment which are fixed supports for the sails, while under the term 'running rigging' are grouped another series of adjustable equipment which have the function of continuously adapting the sail configuration to the actual navigation conditions. From the structural point of view standing rigging can be considered as a unique, balanced system which has to resist the loads transmitted by sails: mast and spreaders resist compression, stays and upwind shrouds resist tension. Chainplates should be dimensioned to resist the higher tension transmitted by shrouds while the attachment points of chainplates to the hull should be reinforced in order to distribute the stresses on side shell and frames. The horizontal component of lower shroud tension also induces compression on deck beams. Similarly the stays should be secured to adequately reinforced attachment points on bow and poop; in this case the vertical component of stays tension induces a considerable bending moment on the hull which must be considered in the scantling of the hull longitudinal modulus. It has been proved in a past investigation on an America's Cup 12 m that the rigging pretensioning alone could cause a hull displacement amidship of about 6 mm (Boote *et al*, 1985).

The mast is subjected to stresses typical of beam-columns, being loaded by axial and lateral forces acting at the same time both in the longitudinal and transverse boat plans. The bending is balanced by stays in the fore-and-aft direction and by shrouds athwartship. Stays and shrouds also help the mast to withstand higher compression loads shortening the unsupported length. Even if the mast resistance should be evaluated globally the study of buckling of the mast is traditionally faced separately in the longitudinal and transverse directions for the sake of simplicity. In the transverse direction the scheme to be considered is that shown in Figure 10.3; it is assumed that the buckled shape of the mast has points of inflexion at the spreaders, that is, the shrouds divide the total height of the mast into panels of reduced lengths. Each panel is considered simply supported owing to the presence of shrouds. The mast is considered fixed at the lower end at deck level and pinned at the upper extremity. With these hypotheses, after having computed the distribution of bending and compression along the mast, the Eulerian buckling loads are computed for each panel adopting the appropriate end conditions.

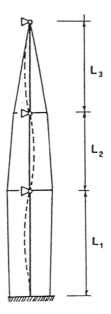

Figure 10.3 *Mast with two sets of spreaders: buckled shape.*

In the longitudinal plan, the task of supporting the mast and of preventing buckling is performed by stays. In this case, however, the considerable length of stays allows major displacements of the nodes, which cannot be considered perfectly supported. Several hypotheses can be assumed about the mast buckled shape in the fore-and-aft direction.

1. Only the headstay and the permanent backstay provide supports to the mast (Figure 10.4a). The unsupported panel length L must be set equal to the total mast height. A clamp and a pin represent the end conditions. This is the most severe condition and leads to very high longitudinal moment of inertia of the mast section; it is however the suitable design condition for cruising yachts.
2. The total height L of the mast is divided into two sections L_1 and L_2 by running backstays and the forestay (Figure 10.4b). This pattern is generally accepted by naval architects as the design condition because it allows the minimum required moment of inertia of the mast section in the longitudinal direction to be substantially reduced. Nevertheless this hypothesis assumes the crew's ability to rapidly trim running backstays when the boat tacks, without leaving the mast unsupported; an error in such an operation can lead to serious trouble in severe sea states. It is then very important to consider this configuration carefully especially for cruising yachts or in all those cases where a high safety level of rigs is required.
3. The mast is divided into two panels by lower shrouds (Figure 10.4c). This can be an hazardous hypothesis because the elasticity of ropes, the small spread

angle of shrouds of the lower spreader does not allow the upper extremity of panel L_1 to be considered to be effectively pinned. Nevertheless, it is beyond doubt that shrouds give some contribution to the mast strength with regard to buckling, even if it is very difficult to quantify it.

4. The mast is divided into three panels (Figure 10.4d) by the running backstays, the forestay and the lower shrouds. The longitudinal schematisation for buckling is equivalent to that athwartship. This is a very optimistic configuration and can be taken into consideration only for pure racing yachts.

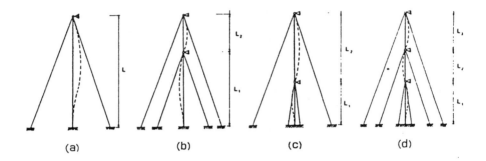

Figure 10.4 *Buckled shapes of mast in the longitudinal direction: (a) mast with forestay and backstay; (b) mast with babystay and running stays; (c) mast with fore and aft lower shrouds; (d) total supported mast.*

It is a matter of course that the case (a) is the most conservative condition and the case (b) is the closest to the real condition; an acceptable compromise comes from considering an intermediate situation between cases (a) and (b), even if the contribution of stays and lower shrouds cannot be neglected. It should be noted that the buckling analysis is usually performed considering the total compressive load applied at the upper extremity of the mast, while, in the real situation, the axial loads are different on each panel. Moreover the designer, however, has to keep in mind that the shrouds, being wires, do not really behave as fixed supports allowing displacements of the nodes in the horizontal direction. To take into account the shrouds contribution and the correct distribution of the compressive loads, it would be advisable to perform a non-linear finite element analysis.

10.4 MAST ARRANGEMENT

Modern masts are all built in aluminium alloys, generally 6000 series, with magnesium and silicon that give to the material high mechanical characteristics and good resistance to corrosion in the marine environment. In the field of sailing yachts the most used alloys are 6063 type for short, economical masts, 6061 type for high quality masts and 6082 type, which is the most expensive one, for racing yacht masts. For very high performance applications also other aluminium alloys like 7000 and 2000 series are employed together with composite materials with carbon reinforcement. Expecially for carbon fibre reinforcements the cost of the material and of the construction procedure keep them out of the normal field of application.

Mast profiles are obtained by extrusion in a wide variety of section shapes in which the inertia in the longitudinal direction (around its transverse principal axis Y–Y) is much greater than that in the transverse direction (around X–X axis).

Table 10.1 Aluminium alloys adopted for sailing yachts spars.

Aluminium alloy	$\sigma_{0.2\%}$ (N/mm^2)	σ_U (N/mm^2)	ε_U	HB	E (N/mm^2)	γ (N/dm^3)
AA 6063	150	195	12%	80	69000	26.5
AA 6061	235	255	8%	80	69000	26.5
AA 6082	255	305	10%	90	69000	26.5

As it is well known the mast induces a vortex zone on the mainsail leading edge, the mast projected area, with a consequent reduction of the sail efficiency. In addition the requirement for higher sailing yacht performance has led to higher aspect ratio of rigs and, as a consequence, the demand of longitudinal inertia is even more greater than the transverse one. An imperative requirement for a mast section is then to provide adequate inertia with minimum dimensions in order to assure good buckling resistance and to avoid as much as possible any interaction with the mainsail.

The section shapes can be divided into three main classes:

- Oval sections
  - used for small–medium size cruising yachts without particular performance requirements. The ratio between the two diameters is about 1.5 while the ratio J_{YY}/J_{XX} ranges between 1.8 and 1.9.
- Bullet sections
  - (or «D» sections) are employed for high efficiency rigs. Even if they are particularly suited for medium–large size yachts, nowadays also very slender profiles are available for little boats. The ratio between the two diameters is about 1.6–1.9 and the J_{YY}/J_{XX} ratio for these types ranges from 2.5–3.
- Open sections
  - are used when a mainsail reefing system is to be set up. The ratio between the two diameters is about 1.8–2.0 while the J_{YY}/J_{XX} ratio ranges between 2.5–2.8. These types of mast are characterised by a high resistance towards compressive loads acting in section plane due to the longitudinal walls and stiffeners. As a drawback the quasi-rectangular shape makes these sections aerodynamically poor.

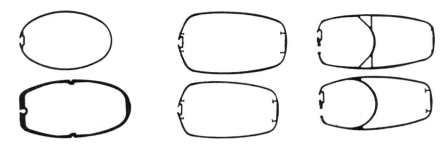

Figure 10.5 *Mast profile section shapes: oval, bullet, rectangular and open.*

While some time ago the oval sections were completely smooth and the sail tracks were externally bolted, nowadays all ovals and bullet sections have built in luff tracks both for bolt rope and for sail sliders; in Figure 10.5 some sections of the three topologies described, of similar dimensions, are reproduced. As an example, for some yachts of different dimensions, the main characteristics of the mast are reported in Table 10.2. All profiles have constant sections and they are provided by the manufacturer in segments of maximum 16–18 m for transport restrictions. As a matter of fact the sections requirements are different at the base and at the top of the mast. At the base the higher compression takes place and a high inertia is necessary while, at the top, it is convenient to reduce weight to the advantage of vessel stability. The majority of spars have constant section along their length; in the case of big and/or high performance yachts it is a common practice to reinforce the mast base and to taper the top. The first is achieved by bolting aluminium strips inside the fore and aft part of the section to increase longitudinal inertia; the more effective alternative consists in introducing a sleeve inside the mast and bolting them together. The same method is employed to create masts longer than 18 m jointing two profiles. In this case a coupling profile is introduced in the mast for 2–3 diameters in length and the two parts are bolted together (Figure 10.6).

Table 10.2 Mast characteristics of some sailing yachts with masthead arrangement

Length	Displ. (daN)	Sail area (m^2)	Mast height (m)	Dimensions (mm)	Thickness (mm)	J_x (cm^4)	J_y (cm^4)
10.0 m	4800	58	12.90	105 × 160	4	260	480
12.0 m	7000	86	16.00	145 × 220	4	790	1540
14.5 m	12700	139	19.00	185 × 315	5.6	1900	3500
18.0 m	22500	218	22.50	220 × 360	6	4700	8300
24.0 m	47000	254	26.42	240 × 410	6	650	12550

The top of the mast is tapered cutting a strip of material from the side of the profile of increasing width. Then the two edges are welded together obtaining a decreasing section towards the masthead (Figure 10.7). This simple procedure allows a reduction in weight and makes the top of the mast more flexible.

Masts can be stepped up in two different manners: simply based on deck, (deck-stepped mast) or based on keel, passing through the deck (keel stepped mast). The first solution is particularly suitable for small–medium size boats which often have to be trailed or to pass beneath low bridges on channels. In addition, if the mast is stepped on deck, there is more space below and the likelihood of water entrance is avoided. The second solution is advisable mainly for structural reasons, especially for large size or cruising yachts, where high compressive stresses take place. Against this, the mast below deck represents a considerable encumbrance for the cabin layout. Moreover the passage of the mast through deck often is source for a leak allowing water to come into the hull; this is amplified by the movements of the mast during sailing which flexes and pumps continuously. Additionally, the mast itself is not waterproof and water can come into the mast through sheave boxes, halyard exits, loose rivets and then reach the internals of the boat.

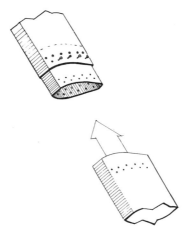

Figure 10.6 *Junction of two mast sections by sleeve.*

200 *Sailing Yacht Design: Theory*

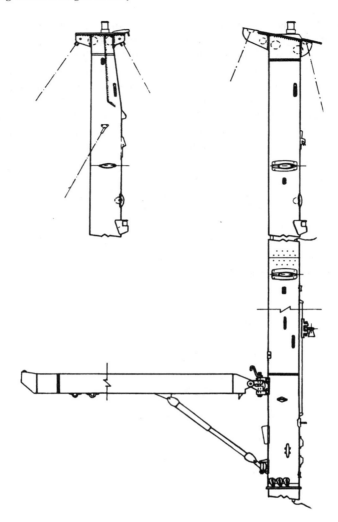

Figure 10.7 *Mast equipment: in particular tapered masthead.*

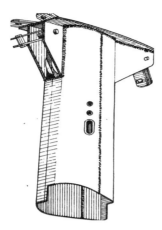

Figure 10.8 *Welded masthead.*

A keel stepped mast is based on the bottom structures, on a stiff longitudinal step which distributes the compression over two or three hull frames. In GRP boats the mast is stepped on a large aluminium or stainless steel plate which has a protuberance exactly matching the inner section profile. The plate is bolted to the rigid bottom structures in correspondence with the keel attachment and then laminated together with the hull shell. The mast passage through the deck is reinforced by two transverse beams and by local longitudinals. The mast is kept in position by wood or plastic chocks; on modern yachts the passage is sealed by big O–rings squeezed vertically between collars bolted to the outer and inner deck shell.

On deck the mast is stepped in a robust tabernacle, hinged by a pivot bolt and blocked by a security retaining pin. Removing the security pin the mast can turn upon the pivot and can be easily unstepped. The base of tabernacle should consist of a large slab bolted to the deck, which distributes the mast compression over a wide deck area. Often partners alone are not sufficient to reinforce the deck to withstand the mast compression, so it becomes necessary to insert a pillar beneath the tabernacle in order to transfer the compression to bottom structure or, better, to reinforce the section with a rigid bulkhead. In these cases the advantage of better space availability below deck is lost; anyway, the solution is mandatory for big sailing yachts that have to pass under bridges.

From the structural point of view the keel-stepped mast has a more efficient constraint at the lower end and the passage through the deck can be considered an additional constraint which absorbs a part of compression. This consideration allows the section and wires dimensions to be reduced saving weight and windage or, as an alternative, the panel span may be increased. Finally the keel-stepped solution allows a hydraulic jack to be used to pump up the spar in racing boats. For the deck stepped mast, the lower panel, which is affected by the higher compression, behaves like a beam column pin jointed at each end. This fact reduces the critical load so a larger mast section should be adopted with, if possible, wider lower shrouds than would be otherwise advisable. It can be said that stepping masts on deck is more difficult and expensive and it results in a larger mast section compared to the keel-stepped solution.

The mast is equipped with a number of fittings, which are connected to the profile wall by bolts, screws, rivets or brackets. All these fastenings require holes in the mast or the use of welding which, reducing the amount of material of the strength section, creating weak zones where collapse might take place. It is then necessary to heed the following advice, also provided by Classification Rules, that:

- Rectangular holes should have rounded corners
- Avoid making more than one hole in the same mast section
- Avoid making holes at the base of the mast where the compression is higher
- If it becomes unavoidable to do so, adopt inserts of appropriate thickness

In the following a brief description of the main fittings on a mast is provided.

Masthead

The forestay and backstay are connected to robust brackets at the top of the mast. The same plates are used to fix tail blocks and sheaves pivots for halyards; mainsail and headsail halyard sheaves are fitted inside the top of the mast section. In the case of masts of a certain importance all these components are usually fitted together on a separate part called 'masthead' which is separately manufactured and then assembled to the mast (Figure 10.8). The masthead can be welded or bolted to the mast top. The first solution has the advantage of a good mechanical connection and lightweight while the second solution allows the masthead to be dismantled for maintenance; this may be necessary, as an example, where a mainsail furling system is installed. The geometry of the masthead should be designed to reduce as much as possible the bending caused by stays; in this respect the attachment point of forestays and backstays should be positioned in such a way that their action lines meet each other on the mast centreline.

Spreaders and shroud tangs

Spreader design is of great importance for the general behaviour of the mast both structurally and aerodynamically. Spreaders are mainly loaded by pure compression transmitted by the shroud tension; in addition mast deformation can cause a bending moment in the horizontal plane. For this reason spreaders are made from aerofoil profiles characterised by a higher inertia in the fore and aft direction; this also contributes to better windage qualities. When the mast needs to be more supported in the longitudinal plane spreaders are angled towards the stern; in this case the bending effect increases and larger sections are needed. This type of arrangement can cause a denting of the mast wall owing to the pushing action of the trailing edge of the spreader.

Spreaders cut the total mast height into panels of reduced length and their efficiency depends on the method of junction to the mast. Spreaders can be fixed to the mast by two different methods. The most reliable solution is represented by a stainless steel or plastic collar, bolted to the mast with two brackets for spreaders accommodation; this solution presents a high strength at the spreader root and prevents the mast-denting phenomenon mentioned earlier. The alternative solution consists of a through bar to which spreaders are connected. This solution is lighter and more streamlined and is particularly advisable for racing yachts. On the other hand it requires a hole of considerable dimensions on the mast profile which reduces the section inertia. This can be dangerous especially on the lower spreaders where compression is higher.

Spreaders should be positioned in such a way as to withstand only compressive loads; bending, associated with fatigue, can lead to the spreader collapse and, as a consequence, to the sudden loss of the corresponding mast support. This causes an increase of the mast panel length and, without an immediate intervention on the rig configuration, buckling takes place. For this reason it is a common practice, also recommended by Classification Rules, to set the spreaders so that their axes bisect the angle of shrouds.

The junction of shrouds to the mast (shroud tang) are mostly realised by bolting two plates into the mast profile connected by a through bar. The shroud is then connected allowing an adequate free articulation; this can avoid fatigue effects which can be dangerous for the shroud and for the mast wall. Another solution consists in exploiting the spreader attachment points for shroud attachment.

The boom

The task of the boom is to trim the mainsail in order to match the optimum attack angle towards the wind. The boom is subjected to bending forces deriving from the tension of the mainsail leech counteracted by the sheet and the vang (kicking strap). The bending moment can be considerably increased when the mainsheet attachment is forward of the leech position. Even if bending acts both in the vertical and horizontal plane, the first one is considerably higher. For this reason boom sections have different inertia in the two main directions of the strength section. The boom is also subjected to compressive loads due to the axial component of clew and vang load. The boom transmits vertical and horizontal forces to the mast through the gooseneck, in a zone where the mast experiences high compressive loads. There are three main type of goosenecks:

- The sliding gooseneck has been adopted for many years on small yachts and it allows tensioning of the luff of the mainsail.
- The fixed gooseneck is more resistant than the previous one. The tensioning of the mainsail luff is achieved by a Cunningham. The gooseneck is fixed to the mast by a bolted stainless steel plate matching the aft profile of the mast.
- Rotating goosenecks are fitted when the boom can be subjected to high torsion loads. Also in this case the gooseneck is fixed to the mast by a bolted plate.

Another fitting which should be considered with attention for the loads it transmits to the mast is the kicking strap (boom vang). Different types exist depending on the boat sizes and employment; starting from the simple block and tackle, lever and rod up to hydraulic units. All of them are connected to the mast by bolted steel or aluminium plates in an area where the mast is subjected to maximum compression.

10.5 DESIGN METHODS

For the calculation of mast and rigging, several simple methods exist in the literature and are proposed by Classification Societies. All of them are based on the determination of the maximum compression on the mast as a function of the boat righting moment. In the following a brief description of some of these methods is presented.

Skene's method

A simplified method to evaluate the maximum compressive load on the mast and the tensile loads on shrouds is known as Skene's method. The maximum compressive load P at the basis of mast is computed from the righting moment of the hull at 30° heel.

$$P = 1.85 P_T \quad \text{where} \quad P_T = \frac{1.5 RM_{30}}{b/2} \qquad [10.1]$$

RM_{30} = righting moment at 30° heel
$b/2$ = lower shroud spreading (Figure 10.9)
1.5 = coefficient taking into account heels greater than 30°
1.85 = coefficient for stays, sheeting and halyard loads

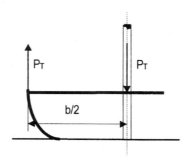

Figure 10.9 *Lower shroud spreading.*

The minimum required moment of inertia of the mast section is given by the following formulas using [Kg] and [cm]:

$$J_{MIN} = 1.422 \ 10^{-7} \ C \ P \ L^2 \ (cm)^4$$

coefficient C is reported in Table 10.3.

Table 10.3 Values of coefficient C

Panels	Mast step	Material	Athwartship		Longitudinal		
			Single spreader	2/3 sets of spreaders	Masthead rig		7/8 fore-triangle
					Short	Tall	
Lower panels	Keel stepped	Spruce	93.3	111.6	53.7	57.8	51.5
		Aluminium	12.9	15.1	7.6	8.0	7.2
	Deck stepped	Spruce	117.8	140.3	58.2	68.0	64.7
		Aluminium	16.2	19.4	8.3	9.4	8.9
Upper panels		Spruce	150.2	167.9			
		Aluminium	20.9	23.3			

If hydrostatic characteristics are not avaliable, the following formula, proposed by Gerritsma (1993), can be useful to obtain a preliminary value of the righting moment:

$$M_R(\varphi) = \Delta \left[\left(0.664 T_c + 0.111 \frac{B_{WL}^2}{T_c} - G_z \right) \sin \varphi + L_{WL} \left(D_2 \varphi F_n + D_3 \varphi^2 \right) \right] \quad [10.2]$$

Δ = displacement of the vessel
L_{WL} = waterline length
T_C = draught of canoe body
φ = heeling angle in radians

$$F_n = \frac{V}{\sqrt{g L_{WL}}} \quad \text{Froude number}$$

$$D_2 = -0.0406 + 0.0109 \left(\frac{B_{WL}}{T_c} \right) - 0.00105 \left(\frac{B_{WL}}{T_c} \right)^2$$

$$D_3 = 0.0636 - 0.0196\left(\frac{B_{WL}}{T_c}\right)$$

For the scantling of shrouds the load distribution of Figure 10.10, derived from experimental measurements, can be assumed (values are vertical components as a percentage of maximum compressive load P_T).

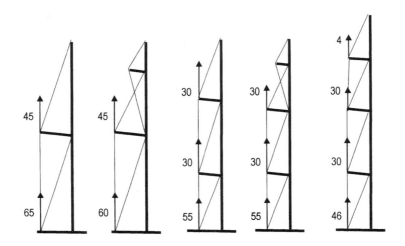

Figure 10.10 *Vertical components of shroud loads as a percentage of compressive load P_T.*

Bureau Veritas method

In the 'Rules for Classification of Yachts' of the Bureau Veritas (1993) there is a section dedicated to spars and standing rigging scantlings. In the following a synthesis of the part regarding masts is reported.

Unstayed mast

The section modulus of the mast at the deck level should be not less than:

$$Z_o = \frac{1.8}{\sigma_y} RM \left[\frac{c' + 0.4P}{a}\right] (cm^3) \qquad [10.3]$$

σ_y = yield stress of mast material (N/mm^2)
RM = maximum righting moment of the yacht in the full load condition (Nm)
c' = vertical distance between the boom gooseneck on the mast and the upper side support through the deck as defined in Figure 10.11 (m)

P = length of the mainsail luff in m (Figure 10.12)
a = design lever arm of the mainsail as defined in Figure 10.12 (m)

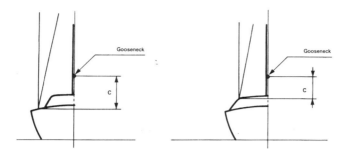

Figure 10.11 *Distance c between gooseneck and chainplate connection point.*

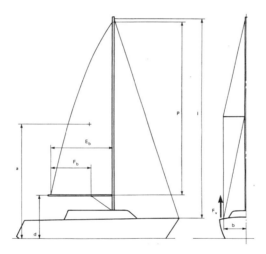

Figure 10.12 *Sail plan symbols.*

Above the gooseneck, the section modulus can be reduced according to the following formula:

$$Z_1 = Z_0 \left[0.2 + 0.8 \frac{h_1}{h} \right] (cm^3) \qquad [10.4]$$

h = length of the mast from the deck support [m]
h_1 = length of the mast between the considered section and the top [m]

208 *Sailing Yacht Design: Theory*

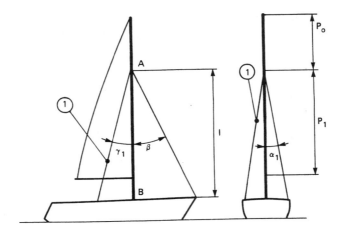

Figure 10.13 *Type I, simple stayed mast.*

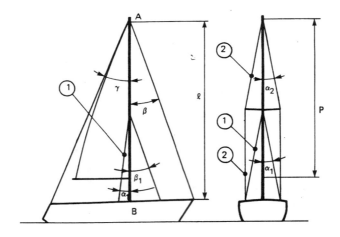

Figure 10.14 *Type I–1, single spreader, mast head.*

Mast and Rigging Design 209

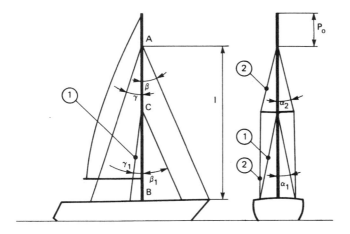

Figure 10.15 *Type I–2, single spreader, 7/8 or 5/6.*

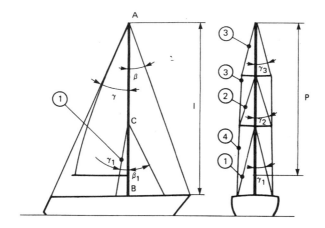

Figure 10.16 *Type II–1, double spreader, mast head.*

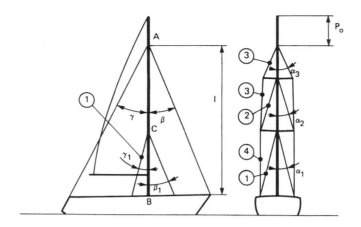

Figure 10.17 *Type II–2, double spreader, 7/8 or 5/6.*

Stayed mast

For stayed mast the method of the BV is based on the determination of the minimum cross-section inertia of mast panels. Different values are provided for longitudinal inertia IL and transverse inertia IT:

$$IT = QF^2 K_2 10^{-4} \qquad IL = QL^2 K_1 K_3 10^{-4} \ (cm^4) \qquad [10.5]$$

$$Q = \left[\frac{RM(P+c)}{b(P+d)} + \frac{1}{3} R \frac{\sin(\gamma + \beta)}{\sin \gamma} \right]$$

β, γ = fore angle of the upper jibstay and after angle of the corresponding backstay
R = minimum breaking strength of the upper jibstay [N]
L = free length of the mast [m]
K_1 = step condition factor (1 for keel stepped mast, 1.3 for deck stepped mast)
K_2, K_3 = as defined in Table 10.4 and Figure 10.16
b,c,d see Figure 10.12.

When no curves of righting moment are available, RM may be determined as follows for yachts with a displacement greater than 0.5 tonne.

$$RM = 1.5 RM_{30} \frac{\Delta}{\Delta_0} + M \qquad [10.6]$$

$$RM_{30} = \frac{4.5\Delta B_{WL}^3 L_{WL}}{100 W_0} + \frac{5.5 W_0^2 T_K}{\Delta} \quad [Nm] \qquad [10.7]$$

Δ_o = light ship displacement [t]
Δ = full load displacement, including crew [t]
W_o = keel weight [t]
L_{wl} = waterline length [m]
B_{wl} = waterline breadth [m]
T_k = draught at the bottom of keel [m]
n = number of persons allowed on board
M = 735b(n + 0.5) heeling moment due to the location of n persons on the gunwale

Table 10.4 K_2 values

Type	Lower panel	Other panels
I	1.7 K_1	–
I–1	1.4 K_1	1.8
I–2	1.4 K_1	1.8
II–1	1.65 K_1	1.8
II–2	1.65 K_1	1.8

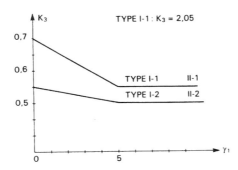

Figure 10.18 K_3 values.

Boom scantlings

The same method also provides a rule for boom scantling; the minimum vertical flexural modulus of the boom section is given by the following relationship:

$$Z_y = \frac{49 F_b E_b P}{\sigma_y} \qquad [10.8]$$

P = length of the mainsail luff in m
E_b = length of the mainsail foot on the boom in m
F_b = horizontal distance between mainsail clew and kicking strap point in m
σ_y = yield stress of the material in N/mm^2

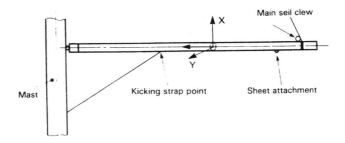

Figure 10.19 *Boom arrangement.*

The section modulus in the horizontal plane Z_x should be not less than 0.4 Z_y.

Spreaders

Spreaders must bisect the angle made by the shrouds (Figure 10.20). The minimum cross section inertia of the spreaders at mid-span is to be not less than:

$$IS = 2.3\ s^2 K_4 R \sin\alpha\, 10^{-4}\ [cm^4] \qquad [10.9]$$

s = length of the spreader [m]
K_4 = 1 if the spreader is simply supported at the mast
K_4 = 0.75 if the spreader is fixed to the mast
R = breaking strength of the relevant shroud [N]
α = angle of the relevant shroud [deg]

The minimum cross-section area, to withstand compression, should be more than:

$$A_s = 1.6 \frac{R}{\sigma_y} \sin\alpha\ [mm^2] \qquad [10.10]$$

σ_y = the yield strength of the spreader material in N/mm^2

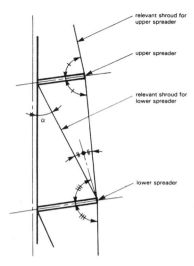

Figure 10.20 *Spreaders arrangement.*

10.6 OTHER CLASSIFICATION SOCIETY RULES

In the construction rules of other classification societies, as for instance in the 'Rules for Construction of vessels less than 15 Metres' (1983) of Det Norske Veritas, there are rules for mast scantling determination that follow the same general approach as illustrated for the Bureau Veritas rules. These methods are based on simplified models of structural behaviour coupled to certain assumptions as to loadings and allowable stresses, tempered by a knowledge of mast failures in service.

To work effectively, each set of rules should be applied as a complete package and no attempt should be made to alter the details. Where a yacht is to satisfy the rules no departure can be made, but in other cases a designer may be tempted to change the rule details to be more 'correct' or 'rational'. Such a temptation should be resisted.

10.7 CONCLUSIONS

The design of mast and rigging of cruising and racing sailboats requires a thorough knowledge and understanding of the complete system in terms of construction solutions, boat management, structural principles and design methods. In this paper the main aspects of such matters have been illustrated. First the principles of how a mast behaves have been explained; then a brief description of mast construction and fitting has been presented. After an assessment of the structural problems regarding masts, the design methods commonly adopted in the spar design have been reviewed; particular attention has been devoted to the rules of Bureau Veritas.

Nevertheless, in the case of big sailing yachts, the previously described methods can cease to be applicable. An incorrect scantling can lead, on the one hand, to excessive dimensions of the mast section, which will cause a decrease in sails efficiency and an increase in hull weight and, on the other hand, to a light dimensioning of the section with dangerous consequences for its strength, especially with regard to buckling. Moreover, in case of races or in severe weather conditions, dismasting is not a rare event, not always ascribable to crew's errors. It becomes therefore suitable to perform a more accurate investigation into the stresses of mast and rigging with finite elements programs which allow simulation of the structure behaviour in a more realistic way, taking into account the tridimensional nature of rigging geometry and loads distribution. Furthermore it is possible to study a number of numerical models corresponding to the sails plans of practical interest under different load conditions. Even though this is laborious for the designer, the final scantlings of the mast can be lighter compared with those obtained by simplified methods.

REFERENCES

1. Boote D., Ruggiero V., Sironi N., Vallicelli A., Finzi B., 1985, 'Stress analysis for light alloy 12 m yacht structures. Comparison between a transverse and a longitudinal structure', S.N.A.M.E., Seventh Chesapeake Sailing Yacht Symposium, Annapolis, U.S.A.
2. Boote, D., 1994, 'Statistical data for pleasure yachts design', (in italian) Tecnica Italiana, Anno LIX, N°4.
3. Bureau Veritas, 1993, 'Rules for the Classification – Certification of Yachts.', Paris, France.
4. Det Norske Veritas, 1983, 'Rules for construction and certification of vessels less than 15 metres.', Hovik, Norway.
5. Gerritsma J., Keuning J.A., Versluis A., 1993, 'Sailing Yachts Performance in Calm Water and in Waves', S.N.A.M.E., Eleventh Chesapeake Sailing Yacht Symposium, Annapolis, U.S.A.
6. Kinney F.S., 'Skene's elements of yacht design.', Adam & Charles Black, London, 1962.

CHAPTER 11
HULL DESIGN AND GEOMETRY DEFINITION

J. Cross-Whiter
University of Southampton

11.1 GEOMETRY OF HULL LINES

The traditional hull lines plan is the complete definition of a complex, three-dimensional surface onto a two-dimensional medium, i.e. paper. In the past, this was the only method of transferring data between designer and builder, calculating hydrostatic quantities, etc. With the advent of full three-dimensional computer-aided draughting the need for two-dimensional paper drawings has been reduced, but the traditional lines plan projections are often still the most useful method for transferring information for manufacturing, hydrostatics, etc.

The lines plan is described in all introductory texts in naval architecture.[1-3] It consists of three orthogonal views, which show the projections of the intersections of the hull with three mutually perpendicular planes. Figure 11.1 shows a lines plan of a 10.9 m sailing yacht.

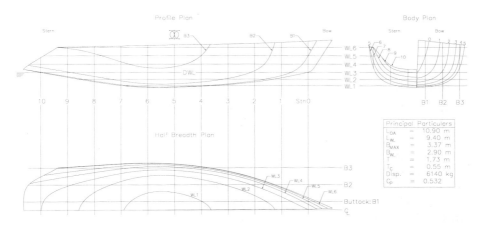

Figure 11.1 *Lines plan of a yacht.*

215

The body plan shows the intersections of the hull with the sections, which are vertical planes transverse to the centre line. The half-breadth shows the waterlines, which are the intersections of the hull with the horizontal waterplanes. The profile or shear plan shows the hull intersections with the buttocks, which are vertical and parallel to the centreplane. Each view in the lines plan displays the shapes of one set of planes, using the edge projections of the other two sets of planes as a reference grid.

Traditionally, sections divide the hull into a number, e.g. 10, of evenly spaced stations, with half or quarter stations defined if necessary near the ends. Waterlines and buttocks are evenly spaced at some appropriate interval, sometimes with smaller intervals near regions of high curvature.

Other plane curves are also often drawn through the hull, for manufacturing purposes or to check fairness in other planes. For example, sections can be rotated relative to either the centreplane or baseplane to show the shapes of cant or inclined frames in the hull. Diagonals are drawn perpendicular to the sections and at some angle relative to the waterlines.

Each set of planes in the lines plan completely describe the hull surface. Therefore, the projections of these planes in each of the orthogonal views must describe the same points in space. This is one component of hull fairness – the agreement in all views about the precise location of points on the hull surface. The second component is a smooth and continuous surface, with no unintended discontinuities in curvature and a minimum number of inflection points.

The process of creating fair hulls in traditional draughting consists of, first, laying down a set of primary definition lines, such as keel, deck, midship section, etc. Then a sparse set of planar curves, such as sections, can be drawn to define the hull surface. Points picked off of these curves then define preliminary waterlines and buttocks. When any of these curves are faired to the desired shape, this alters the shapes of the other curves, which are then refaired and so on, until the process spirals in towards a finished design.

11.2 COMPUTER FAIRING

General purpose CAD packages

General purpose CAD packages are not specifically designed to fair hull lines, but are nevertheless often used for drawing lines. In a conventional draughting package, as opposed to a solid or surface modelling package, each curve in the drawing is an independent entity, so it is up to the designer to ensure that all of the curves define the same surface. When using these packages in 2D mode the process is much like traditional paper-based lines fairing, with three independent views, each using two-dimensional polylines for the ship curves. As in paper-based fairing the coordinates in each view must be made to agree to ensure a unique surface. Lines can also be faired in three dimensions, which saves having to draw three views, but it is still up to the designer to ensure that the curves define a unique surface.

Hull Design and Geometry Definition 217

Solid and surface modelling packages generate unique, well-defined surfaces, often employing the same techniques as the special purpose lines fairing software discussed below. Relative to the general mechanical engineering surfaces these packages are designed to represent, ship hulls are relatively simple, and can easily be defined. However this type of software does not necessarily include all of the essential tools for the naval architect, such as curvature display, hydrostatics information, and cut planes through the surface. The more sophisticated packages, such as those used in the automotive industry, provide more of these tools, but are often more expensive and difficult to use than special purpose lines fairing software.

Lines fairing software

In computer lines fairing the traditional lines fairing process is in some senses reversed. Orthogonal lines are not necessarily used to define the hull surface, but once the surface is defined, sections, waterlines and buttocks are defined by intersecting the appropriate planes with the hull surface.

Computer lines fairing software mathematically defines the ship hull as either a grid of intersecting curves on the surface or as a surface defined by a grid of intersecting polygonal control lines. Each curve intersection is uniquely defined in space, thus giving a uniquely defined hull surface. Therefore one of the time consuming parts of traditional fairing, altering one set of curves to reflect alterations made to another set of curves, is eliminated. Figure 11.2 shows the network curves used to define the lines plan of Figure 11.1.

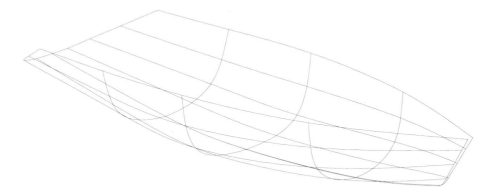

Figure 11.2 *Network curves for a yacht.*

There are a number of algorithms available for defining hull surfaces, but the two most common are splined curves on the hull surface and surface definition from a network of off-body control polygons.

The algorithms used to define these surfaces are outlined in a number of computer draughting and differential geometry texts.[4-8]

Curves and intersections

To understand how a hull surface is built up of intersecting curves or control lines it is helpful to first look at the definition of a single curve in space.

The curve is defined by series of X,Y,Z coordinates. In lines fairing software the curve between defined nodes is usually interpolated by spline functions. A commonly used spline is the cubic spline, which describes an elastic beam held at the nodes, analogous to a draughtsman's spline held by weights. This curve has an optimum smoothness property, i.e. minimum RMS curvature, analogous to the minimum strain energy of the elastic beam.

Between any two nodes the curve is described by a cubic polynomial of the form:

$$X = a_o + a_1 u + a_2 u^2 + a_3 u^3 \tag{11.1}$$

a_n = constant coefficients
u = a parameter along the curve

Splines can be generated that interpolate one coordinate with another, e.g. Y as a function of X. However, in this form the spline does not allow vertical tangents. A more practical form is the parametric spline, in which X, Y, and Z are splined independently against some parameter, u, allowing 0 slopes and curvatures in any of the coordinates. If u runs from 0 to 1 in each interval, the spline is a uniform spline. A non-uniform spline is generated by allowing different ranges of u within each interval. For example, u can run from 0 to L, the straight line distance between the nodes.

Within each interval from u_i to u_{i+1} taking the first and second derivatives of Equation 11.1 and substituting the results back into Equation 11.1 yields:

$$X = X_i + \dot{X}_i u + \tfrac{1}{2}\ddot{X}_i u^2 + \frac{\ddot{X}_{i+1} - \ddot{X}_i}{6 L_i} u^3 \tag{11.2}$$

$\dot{X}_i$ and $\ddot{X}_i$ refer to the slope and curvature, respectively, at the ith points
$L_i = u_{i+1} - u_i$.

Further substitutions yield:

$$\begin{aligned}\dot{X}_i &= \frac{X_{i+1} - X_i}{L_i} - \frac{L_i}{6}(\ddot{X}_{i+1} + 2\ddot{X}_i) \\ \dot{X}_{i+1} &= \frac{X_{i+1} - X_i}{L_i} + \frac{L_i}{6}(2\ddot{X}_{i+1} + \ddot{X}_i)\end{aligned} \tag{11.3}$$

Slope continuity is required at each point, so the slope at point i given by the first part of Equation 11.3 over the interval u_i to u_{i+1} must equal that given by the second part of Equation 11.3 over the interval u_{i-1} to u_i:

$$L_{i-1}\ddot{X}_{i-1} + 2(L_i + L_{i-1})\ddot{X}_i + L_i \ddot{X}_{i+1} = \frac{6}{L_i}(X_{i+1} - X_i) - \frac{6}{L_{i-1}}(X_i - X_{i-1}) \tag{11.4}$$

Hull Design and Geometry Definition 219

Given n points this equation requires the value X at n points, plus 2 other pieces of information, to solve for the unknown $\ddot{X}$ values. The additional information can be the slope or curvature at the two ends of the curve. Given the curvature at the curve ends a system of linear equations based on Equation 11.4 can be solved for the $\ddot{X}$ values.

If each coordinate is independently splined against the curve parameter u each component of the slope or curvature end condition can be independently set, allowing control over both the direction and magnitude of the end condition vector. For example, if set slopes are used, this allows control over slope tangency and fullness at the ends.

Cubic spline curves can be faired by moving the node locations or by altering the curvature values at each node. For the latter, a curvature influence function is established, which gives the change in X for all points, given a change in $\ddot{X}$ at one point.

For interactive curve design many people are uncomfortable with the requirement for additional information at the ends of the curve, in addition to the positional information. This has motivated spline formulations that require only positional information along the curve. The two best known examples of this type of formulation are the Bezier and B-Spline curves. These curves are generated from the positions of vertices of a control polygon, rather than points on the curve itself.

A Bezier curve of nth degree is defined by:

$$X(u) = \sum_{i=0}^{n} \frac{n!}{i!(n-i)!} u^i (1-u)^{n-i} P_i \quad 0 < u < 1 \qquad [11.5]$$

P_i = the position of the ith control vertex

For example, a third degree curve defined by points 0 through 3 is given by:

$$X(u) = (1-u)^3 X_o + 3u(1-u)^2 X_1 + 3u^2(1-u) X_2 + u^3 X_3 \qquad [11.6]$$

Similar equations determine the Y and Z values.

The degree of the curve determines the number of vertices required to define it. A single segment is defined by $n + 1$ vertices. Unless the interior points are colinear, the curve does not pass through them. However, by taking the first derivative of Equation 11.6 it can be seen that the curve passes through, and is tangent to, the control polygon at the ends. Composite curves with more than $n + 1$ vertices are defined by splitting the curve into a number of overlapping polygons. To enforce slope continuity at the ends of these polygons the last segment in each must be colinear with the first segment in the following polygon.

Another spline which uses a control polygon to define the curve is the B-spline, used in many hull drawing packages.

In this spline the points along an kth order curve are given by:

$$X(u) = \sum_{i=1}^{n+1} P_i N_{i,k}(u) \qquad [11.7]$$

P_i = the position of the ith control vertex
$N_{i,k}$ = a basis function defined by the recursion formula:

$$N_{i,1}(u) = \begin{cases} 1 & \text{if } t_i < u < t_{i+1} \\ 0 & \text{otherwise} \end{cases}$$

$$N_{i,k}(u) = \frac{(u-t_i)N_{i,k-1}(u)}{t_{i+k-1}-t_i} + \frac{(t_{i+k}-u)N_{i+1,k-1}(u)}{t_{i+k}-t_{i+1}}$$

[11.8]

The variable *t* refers to a parameterization vector, called a knot. In a uniform B-Spline the knot vectors have integer values. Typically these increase by one for each control point and range from 0 to $n - k + 2$ if there are $n + 1$ separate control points in the polygon. $n + 1 + k$ elements are required in the knot vector, so knots beyond the ends of the control polygon are required to define the end segments. Typically, multiple knots are used, such that $t = 0$ for i from 1 to k, and vertices with $t = n - k + 2$ are added for $i > n + 1$.

Discontinuities are introduced in the curve by multiple knots, that is repetitions of knot vectors. $k - 1$ identical knots introduces a discontinuity in a kth order curve. For example a double knot gives a discontinuity in a third order curve, a triple knot is discontinuous on a fourth order curve, etc.

The knot vectors do not have to be integers, but can be any parameterization along the curve, such as chord length between control vertices, as discussed above for cubic splines. This produces a non-uniform B-spline.

An additional weighting can be applied to each basis function defined in Equation 11.8, such that:

$$X(u) = \frac{\sum_{i=1}^{n+1} P_i N_{i,k} W_i}{\sum_{i=1}^{n+1} N_{i,k} W_i}$$

[11.9]

This produces a rational B-spline.

Bezier and B-spline curves are thus weighted averages of the locations of the vertices composing the control polygon. As such, they possess the convex hull property, which means that any point on the curve is contained within the bounds of its defining polygon.

Figure 11.3 shows cubic splines and B-splines of various orders drawn using the same control vertices. The control vertices are indicated by the thick straight lines.

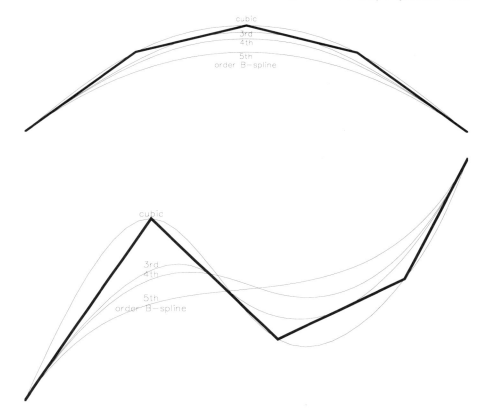

Figure 11.3 *Various spline types on two control polygons.*

Surfaces

Hull surfaces are composed of networks of intersecting cubic, Bezier or B-Spline curves. Generally the network is composed of a series of quadrilateral patches, with curves running primarily girthwise intersecting curves running primarily longitudinally. Each intersection of a longitudinal and girthwise curve uniquely defines a point in space. Editing this location alters both of the curves running through it. Since only 2 sets of mutually orthogonal curves (in a parametric sense) are used to define the hull each intersection point is unique and the hull surface is well defined. Other grid types, including triangular or polygonal patches and intersections between curves of the same type (longitudinal-longitudinal or girthwise-girthwise) have been devised, but most commercial packages continue to use quadrilateral patches.

The extension of spline definition from curve to surface is relatively straightforward.

Within the interval from 0–1 Equation 11.1 can be rewritten as:

$$X(u) = X(0)(1 - 3u^2 + 2u^3) + X(1)(3u^2 - 2u^3) + \dot{X}(0)(u - 2u^2 + u^3) + \dot{X}(1)(-u^2 + u^3)$$
[11.10]

This can be written in matrix form as:

$$X(u) = \begin{bmatrix} 1 & u & u^2 & u^3 \end{bmatrix} \mathbf{F} \begin{bmatrix} X(0) \\ X(1) \\ \dot{X}(0) \\ \dot{X}(1) \end{bmatrix}, \text{ where } \mathbf{F} = \begin{bmatrix} 1 & 0 & 0 & 0 \\ 0 & 0 & 1 & 0 \\ -3 & 3 & -2 & -1 \\ 2 & -2 & 1 & 1 \end{bmatrix}$$
[11.11]

Points on the bicubic patch are defined in terms of the orthogonal parameters u and v. Equation 11.11 is extended into uv space by:

$$X(u,v) = \begin{bmatrix} 1 & u & u^2 & u^3 \end{bmatrix} \mathbf{FQF}^{\mathbf{T}} \begin{bmatrix} 1 \\ v \\ v^2 \\ v^3 \end{bmatrix}$$
[11.12]

where F is above and

$$\mathbf{Q} = \begin{bmatrix} X(0,0) & X(0,1) & X_v(0,0) & X_v(0,1) \\ X(1,0) & X(1,1) & X_v(1,0) & X_v(1,1) \\ X_u(0,0) & X_u(0,1) & X_{uv}(0,0) & X_{uv}(0,1) \\ X_u(1,0) & X_u(1,1) & X_{uv}(1,0) & X_{uv}(1,1) \end{bmatrix}$$
[11.13]

Subscripts refer to first derivatives with respect to the indicated parameter. X_u and X_v are given by the curves along the four borders of the patch. X_{uv}, often referred to as the twist vector, is derived from the u derivative of the v derivative data, or vice versa. It is important in determining the characteristics of the surface between the defined border curves.

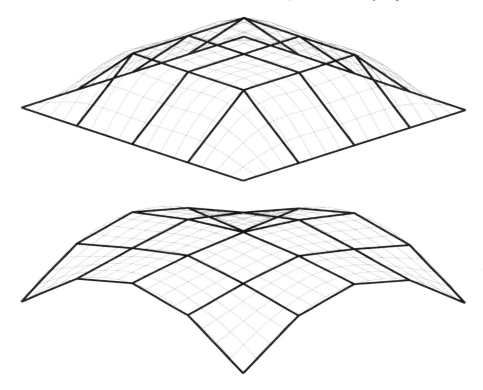

Figure 11.4 *Bicubic surfaces*

Figure 11.4 shows examples of bicubic surfaces defined over two different control networks.

For Bezier surfaces the summation in Equation 11.5 is extended into two dimensions by introducing a second set of basis functions and performing a double summation:

$$X(u,v) = \sum_{i=0}^{m}\sum_{j=0}^{n} J_{m,i}(u) K_{n,j}(v) X_{i,j}, \text{ where}$$

$$J_{m,i}(u) = \frac{m!}{i!(m-i)!} u^i (1-u)^{m-i}$$ [11.14]

$$K_{n,j}(v) = \frac{n!}{j!(n-j)!} v^j (1-v)^{n-j}$$

The special case of a fourth order Bezier patch in the interval $0 \leq u \leq 1$, $0 \leq v \leq 1$ can be expressed in the same form as Equation 11.12, but in this case

$$\mathbf{F} = \begin{bmatrix} 1 & 0 & 0 & 0 \\ -3 & 3 & 0 & 0 \\ 3 & -6 & 3 & 0 \\ -1 & 3 & -3 & 1 \end{bmatrix}, \text{ and } \mathbf{Q} = \begin{bmatrix} X(0,0) & X(0,1) & X(0,2) & X(0,3) \\ X(1,0) & X(1,1) & X(1,2) & X(1,3) \\ X(2,0) & X(2,1) & X(2,2) & X(2,3) \\ X(3,0) & X(3,1) & X(3,2) & X(3,3) \end{bmatrix}$$

[11.15]

So each point on the fourth order Bezier patch is a weighted sum of the vertices of a 16-point control polyhedron.

Similarly a B-Spline patch is defined by the double summation:

$$\sum_{i=1}^{m+1} \sum_{j=1}^{n+1} N_{i,k} M_{j,l} P_{i,j} \qquad [11.16]$$

where the basis functions $N_{i,k}$ and $M_{j,l}$ are defined as in Equation 11.8.

The special case of a fourth order B-spline patch in the interval $0 \leq u \leq 1$, $0 \leq v \leq 1$ can be expressed in the same form as Equation 11.12, but in this case

$$\mathbf{F} = \frac{1}{6}\begin{bmatrix} 1 & 4 & 1 & 0 \\ -3 & 0 & 3 & 0 \\ 3 & -6 & 3 & 0 \\ -1 & 3 & -3 & 1 \end{bmatrix} \text{ and } \mathbf{Q} \text{ is as for the Bezier surface.} \qquad [11.17]$$

More generally, as in Equation 11.9, the knot vectors in the i and j direction can be integers or non-uniform, and the weighting functions can be normalised by a set of weights, $W_{i,j}$ assigned to each point. A surface with non-uniform knot spacing and rational weighting values is known as a non-uniform, rational B-spline surface (NURBS).

If the control vertex spacing varies greatly a uniform knot spacing creates anomalies in the surface. A non-uniform knot vector allows more flexibility in control vertex spacing, but can create discontinuities at the joint between two surfaces if the knot vectors in the two surfaces do not match along the edge.

Rational weights allow the definition of shapes that are impossible with non-rational splines, such as conics, but can create their own set of fairing difficulties and continuity difficulties between surfaces.

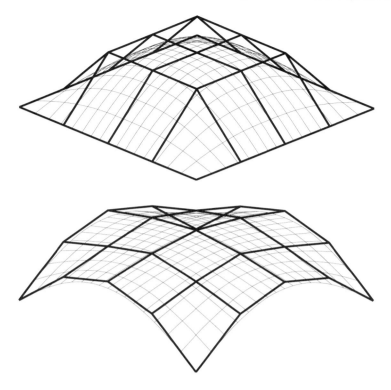

Figure 11.5 *4th order B-spline surfaces*

Figure 11.5 shows two examples of fourth order B-spline surfaces, drawn over the same control network as Figure 11.4. Fourth order B-splines have equivalent flexibility and continuity characteristics as cubic splines.

Surface metrics

A surface defined by one of the methods described above satisfies one definition of fairness, that of a unique definition of each point on the hull surface. However, it still remains up to the designer to ensure the second definition of fairness, that of smoothness. There are algorithms for automatically smoothing curves and surfaces from an input set of points,[9] and some packages provide this facility. For example, one algorithm deletes and re-inserts vertices, and is analogous to the traditional method of lifting a spline weight and allowing the spline to relax to a new faired shape. Regardless of the specific algorithm used, automatic surface fairing involves a compromise between a desired degree of smoothness and an acceptable variation from the original input points.

It is more common to interactively fair the surface to achieve the desired shape. The designer can fair the hull approximately by moving control points, and observing the resultant shapes. However, observing the hull on a computer screen will not reveal all of the subtle changes in curvature that will be noticeable on a finished surface at full scale. Therefore, the software provides indicators of surface metrics, which are measures of surface characteristics, such as curvature. The hull is faired by moving the control points or adjusting curvatures directly, until the desired curvature distribution is achieved.

At each point along a curve a circle can be drawn which is tangent to the curve at that point, and whose radius equals the radius of curvature of the curve. The curvature vector at that point is directed perpendicular to the plane of the circle, and has a magnitude equal to the inverse of circle's radius. For a parametrically defined curve this vector is given by:

$$\mathbf{k}(u) = \frac{\dot{\mathbf{R}}(u) \times \ddot{\mathbf{R}}(u)}{\left|\dot{\mathbf{R}}(u)\right|^3} \qquad [11.18]$$

$\mathbf{R}$ = vector defining a point on the surface

the dotted notation refers to the derivatives of $\mathbf{R}$ with respect to u, i.e.

$$\dot{\mathbf{R}} = \left[\dot{X}(u), \dot{Y}(u), \dot{Z}(u)\right].$$

The magnitude of this vector is the scalar curvature, κ.

At a point on a surface the intersection of the surface and a plane normal to the surface forms a curve with curvature κ, which is a function of the rotation angle of the plane about the surface normal vector $\mathbf{n}$. The curvatures at the angles for which the curvature is a minimum and maximum are the principal curvatures, κ_1 and κ_2, and are the two solutions to the system of equations:

$$\begin{vmatrix} \mathbf{R}_u\mathbf{R}_u\kappa - \mathbf{R}_{uu}\cdot\mathbf{n} & \mathbf{R}_u\mathbf{R}_v\kappa - \mathbf{R}_{uv}\cdot\mathbf{n} \\ \mathbf{R}_u\mathbf{R}_v\kappa - \mathbf{R}_{uv}\cdot\mathbf{n} & \mathbf{R}_v\mathbf{R}_v\kappa - \mathbf{R}_{vv}\cdot\mathbf{n} \end{vmatrix} = 0 \qquad [11.19]$$

where single subscripts refer to partial derivatives of the surface with respect to the indicated parameter, double subscripts refer to second partial derivatives with respect to the indicated parameter(s), i.e. $\mathbf{R}_u = \partial\mathbf{R}/\partial u$, $\mathbf{R}_{uu} = \partial^2\mathbf{R}/\partial u^2$, etc., and $\mathbf{n}$ is the normal vector.

One commonly used metric is the Gaussian curvature, defined as the product of the principal curvatures, and given by

$$K = \kappa_1\kappa_2 = \frac{(\mathbf{R}_{uu}\cdot\mathbf{n})(\mathbf{R}_{vv}\cdot\mathbf{n}) - (\mathbf{R}_{uv}\cdot\mathbf{n})^2}{(\mathbf{R}_u\mathbf{R}_u)(\mathbf{R}_v\mathbf{R}_v) - (\mathbf{R}_u\mathbf{R}_v)^2} \qquad [11.20]$$

Multiple surfaces

Often a complex shape, such as an appended hull or hull with flat of side, is most easily defined by using multiple surfaces. The common edge may be a curve which is defined on both surfaces. For the two surfaces to match exactly the curves on both surfaces must contain identical vertices, and if using NURBS, identical knot vectors and weights. Approximate matches can also be defined, for example in cases where the grids for the two surfaces are not linked, by manually ensuring that the two edge curves match within construction tolerances.

Alternatively, some CAD and lines fairing packages allow trimmed and relational surfaces. In these cases the surface geometries are defined independently, and the relationship between surfaces is defined. Trimmed surfaces intersect, with the overlap trimmed off. A change in one surface does not affect the other, only the trim line is redefined. Relational surfaces are attached, so that alterations in one surface alters the other, maintaining the relationship between the two.

For example a keel defined by a specific planform and section shape may be joined to the bottom of a canoe body. A trimmed surface would allow alterations to the canoe body without changing the keel shape. On the other hand, a feature such as a spray rail on a motor yacht would best be defined by a relational surface, so that it retained its relative position on the hull surface as that surface was altered.

11 3 OTHER SURFACE DEFINITIONS

Generally the spline techniques described previously are the most practical for defining a wide range of hull shapes. However, in certain circumstances other formulations or analytical expressions for known shapes can be the most efficient and accurate.

Developable surfaces

A developable surface is one which can be produced from a flat sheet of material using only bending processes, with no stretching. Such a surface has 0 Gaussian curvature, since the principal curvature is zero in one direction. The curvature in the orthogonal direction represents the amount of bending required to achieve the desired shape. The lines along which the curvature is zero are called generators of the surface. If they are parallel the surface is ruled, and if they intersect at a point the surface is conical. Developed surfaces can be generated directly by projecting a series of straight line generators from a set of defined points, or along a fixed direction. However, it is difficult to distribute the vertices appropriately to obtain the desired surface characteristics. Therefore it is more common to use developable surface techniques in combination with the hull surface algorithms described above to generate developable surfaces from defined boundary curves, as shown in Figure 11.6.

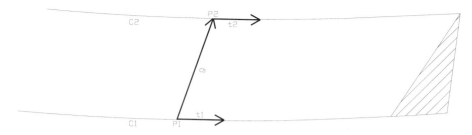

Figure 11.6 *Developable surface generator between two defined curves.*

A generator g from a vertex crosses curve C1 at point P1, and curve C2 at P2. **g** is the vector segment from P1 to P2. **t1** is the tangent vector at P1 and **t2** is the tangent at P2. For a developable surface to exist between these points **g, t1** and **t2** must be coplanar, so

$$\mathbf{g} \times \mathbf{t1} \cdot \mathbf{t2} = 0 \qquad [11.21]$$

Finding the developable surface between the two curves then involves stepping along one curve at a specified interval and, at each point, finding the point on the other curve such that Equation 11.21 is satisfied. If no such points can be found, or generators intersect within the surface boundaries then the surface is not developable, and one or both of the curves must be modified until developability is achieved. Unless the ends of the surface lie along generators, the above stepping procedure will find the end of one curve before the other. In these cases some end condition is imposed to finish off the triangular patch bounded by the end of the patch, the end point of the finished curve and the residual part of the other curve, as shown by the hatched area in Figure 11.6. This patch can be filled by generators radiating from the end point of the finished curve, making that point a generator, or by extending this curve until the end point of the other curve is reached.

Once all of the generators have been found for a surface the curvatures perpendicular to these generators should be checked, to ensure that the intended construction material can be elastically bent to the required curvature.

Sections, waterlines, buttocks, etc. through the hull can be found by searching for the intersections of the straight line generators with these planes.

Finally, the unrolled panel shape for the surface can be determined from the generators. In a given rectangular region bounded by points P1, P2, P3, and P4 the lengths of **P1P2, P2P4, P1P3** and **P3P4** are known. Given the position of segment P1P2 on a flat sheet of material, the positions of the other three segments can be laid out. This provides the position of **P3P4**, which is used as the starting segment for the next rectangular panel, and so forth up to the end of the surface.

Analytical formulations

To avoid the lengthy design spiral there has long been an interest in mathematically defining ship hulls, so that any point on the surface can be defined by equations.[10,11] However, the resulting shapes are not generally useful for yacht or commercial design work, and given the capabilities of modern lines fairing software, this approach is not generally followed.

Parts of a yacht that can often be defined analytically are appendages. In some cases the planform is a quadrilateral, which can be defined by its root chord length and position, tip chord length and position, span and sweep angle. The span is defined as the distance between the root and tip chords, measured perpendicular to the flow direction. The sweep is often given in terms of quarter chord sweep, i.e. the angle of the line joining the points 1/4 of the distance along the root and tip chords, measured from a line parallel to the span. If the leading and trailing edges are not straight lines, they can be defined by splines running through a defined set of points.

Each waterline along the span is a specific aerofoil section shape, defined by the chord length between leading and trailing edges, mean line (if not symmetrical) and thickness. The mean line is the curve that runs from the leading edge to the trailing edge, midway between the two surfaces of the appendage. The maximum offset of this line from the straight line joining the leading and trailing edges is called the camber of the aerofoil, and is generally given as a fraction of the chord length. To define the appendage surface the half thickness at each point along the chord is added perpendicular to the mean line. Near the leading edge the aerofoil is defined by its leading edge radius.

Some of the most common aerofoil shapes are still the NACA sections dating back over 50 years.[12] In that reference the thickness sections are divided into a number of families. The section shapes within each family share basic shape characteristics, such as position of maximum thickness and leading edge radius characteristics, with varying thickness/chord ratio. The shape designation begins with the family identifier, followed by a number specifying the thickness/chord ratio. Commonly used sections are the NACA 00xx family, with relatively blunt leading edges and maximum thickness at 30% chord, and the NACA 64-xxx family, with thinner noses and positions of maximum thickness at 40% chord. Other families, such as the 65 and 66, have thinner noses and/or further aft positions of maximum thickness.

Points along the NACA 00xx sections are defined by the equation;

$$\pm Y_t = t/c/0.20 \cdot \begin{bmatrix} 0.29690(x/c)^{1/2} - 0.12600(x/c) - 0.35160(x/c)^2 \\ + 0.28430(x/c)^3 - 0.10150(x/c)^4 \end{bmatrix} \qquad [11.22]$$

t/c = the thickness/chord ratio
x/c = the fraction of chord from the leading edge

Points for other section shapes are not given analytically, but in tables of offsets of y/c versus x/c. Mean line offsets are given in similar tables.

Since the NACA data were published other authors have published tables of sections for various uses[13]. Generally these do not fall into families in the way that the NACA sections do, and the mean lines are not defined independently of the thickness offsets. Rather, the total offsets of the two foil surfaces are given from a datum line. Often, these sections have been developed for aerodynamic purposes, and are cambered.

11.4 APPROACHES TO DESIGN

When using computer-aided lines fairing packages, there are several different ways of starting the design spiral on a given project. Many of the basic methodologies are similar to those used in traditional lines fairing, with modifications to account for the specific capabilities of the lines fairing package.

Design from chosen dimensions or parameters

When designing from chosen dimensions, the designer starts with a simple definition of certain key dimensions and curves, e.g. the deck line, keel line, midship section, transom and stem. Once these curves are initially defined, the design is refined by editing these curves and adding new curves to the network. Points on these initial curves might be entered numerically, or digitised in from an existing drawing, as described below. Alternatively, since editing points on the surface is easy, the lines fairing software can generate a primitive surface, such as a flat plate or cylinder, from a sparse set of input parameters, such as number of control point rows and columns, length and beam. This surface does not represent an attempt to define a realistic surface, but rather a starting point for subsequent editing.

The form parameters and surface metrics are calculated regularly throughout the design process to check on the progress towards the desired shape. The software facilitates these checks by quickly calculating these quantities, which allows the designer to fine-tune the shape to any desired level of accuracy.

The key concept in this approach is to keep the surface network as simple as possible throughout the process, only adding control points where required to refine the shape definition. A simple network is easier to fair than a complex one. Also, since the planes required for a conventional lines plan can be generated from the defined surface, there is no need to fair using conventional lines plan curves, such as sections or waterlines. Rather, the control grid should define key feature lines through the hull, such as turn of bilge, flat of bottom, diagonals, etc., which may not correspond to conventional lines plan curves.

Digitising

Often, early in the process of switching to a computer-based system from a paper-based one or when reproducing historical hull forms, it is necessary to base a new design on a set of lines already drawn on paper. The offsets can be measured from the drawing and entered numerically into the lines fairing software, but it is much easier to digitise the points in, if the software provides this facility.

In this process, similar guidelines to those outlined above apply, i.e. keep the network as simple as possible, only adding those lines that are essential to defining the hull shape. It is possible to digitise in every station in the lines plan, using large numbers of points along each station, but given the combination of small fairing, lines plan reproduction and digitising errors, the resulting curves become difficult to fair.

In software that uses on-surface splines the digitised points are defined directly on the surface. In B-spline packages, the software computes the positions of control vertices that give the digitised points. This process is more or less accurate, depending on the B-spline formulation used. In any case, even with on-surface splines, control curves that describe the original digitised points should be maintained throughout the fairing process, to provide a check on the deviation of the faired surface from the original. If allowed by the software, these curves should be digitised at a finer resolution than the curves actually used to define the hull.

Adaptation of existing form

On new projects commercial and yacht designers often modify existing designs that have proven successful in the past. This procedure is convenient, as it saves the step of initial network definition, and, provided the alterations are systematic and logical, the amount of refairing required is minimal. From a performance point of view the procedure is also attractive, since small changes to a successful design are likely to be successful. Computer lines fairing packages make such alterations, and the subsequent fine-tuning of the form parameters, relatively quick and easy.

Aside from editing the individual control points on the curve network, the most obvious ways the lines fairing package can facilitate design alterations are by scaling and translating coordinates. All of the specified coordinates are simply multiplied by the scale factor or shifted by the translation factor. Individual coordinates can be scaled independently of the others, to alter the proportions of the hull.

In commercial ships systematic translations of sections along the ships have been used to produce specified shifts in parameters such as LCB and C_p.[14] These types of techniques may not be generally applicable to yachts, but systematic use of the translation, scaling and curve editing facilities, in conjunction with the computed hydrostatics, enables quick and accurate alterations to hull form characteristics.

Given the relative efficiency with which modifications to existing forms can be made, the designer is well advised to maintain an extensive library of parent hull forms.

11.5 LINKS TO OTHER DESIGN SOFTWARE

Ordinarily the definition of a hull surface is not an end in itself but rather a means of providing data for subsequent steps in the design process. Data must be generated by the lines fairing software for hydrostatics calculations, manufacturing, detail design, finite element structural calculations or CFD. For these purposes the software provides various methods of interrogating the surface to generate data in the form required by the other software.

Planar curves

For many purposes the traditional orthogonal lines plan curves are required, even though the surface may have been designed on another set of curves. The lines plan curves are defined by the intersections of defined planes with the hull surface. Similarly, the surface may be trimmed along defined planes or surfaces. The intersection lines are found by stepping the curves along the surface, and searching for intersection points. There are various approaches to this problem but the specifics of the process are not apparent to the user, and will not be discussed here.

Although general purpose CAD packages are often not ideal for fairing hull lines, as discussed previously, they are often used for detail design, using the output of lines fairing software. For example, the lines fairing software can generate the shapes of the intersections of frames with the hull surface. The CAD package can then be used to design the frames, based on these curves. Similarly for decks and superstructures.

Hydrostatics and stability information is often calculated on a section-by-section basis, as this facilitates the integration of sectional quantities at specific waterlines, heel angles, etc. Therefore hydrostatics and stability software requires data hull surface offsets along a series of planar curves.

Dynamic data exchange, particularly within integrated software suites, allows this type of data to be exchanged in real time between different programs, but it is more usual to exchange data in files with commonly agreed formats. Certain file formats have evolved into standards for specific types of software, such as IMS and GHS files for hydrostatics software and DXF and IGES for CAD software. However, software should be able to read and/or write files of various formats to maximise its flexibility. It should be noted that there is often disagreement between programs on the interpretation of even supposedly well defined file formats such as DXF and IGES.

Points on the surface for gridding

CFD and finite element software require a well-defined grid of points on the hull surface, as opposed to a sequence of curves as discussed above. For some purposes the control grid used to define the hull is a good starting point. For example, a hull can be gridded for a panel method CFD program by running a number of isoparamatric curves at specified intervals along the surface, both girthwise and longitudinally. If the control network has longitudinals running along such feature lines as bilge or chine lines, the placement and orientation of the CFD panels roughly correlate with the flow field, improving the numerical stability of the CFD calculations. If a complete hull has been defined up to the deck, care must be taken to remove degeneracies in this type of grid at the free surface.

More sophisticated CFD codes, such as Navier-Stokes solvers, can also use these panels as starting points for generating cells in the fluid domain. However, these solvers have particular gridding requirements, which do not necessarily correlate with the control grid used to fair the hull. Therefore, special gridding software is used, which solves for the grid distribution on the hull surface and uses the hull surface control grid to locate the surface at each CFD grid point. Thus the output grid used in the CFD code may have little relationship to the control grid used to define the surface.

Many finite element and CFD packages link directly with general purpose CAD packages, allowing easy transfer of data between them. However, finite element and CFD packages are sensitive to errors in the surface definition, such as holes, so automatic gridding from CAD data is not always as straightforward as it appears. For example, small errors at the joints between trimmed or relational surfaces can cause gaps or degenerate patches. Similarly, in solid modelling packages, junctions between bodies require a much finer resolution than the bodies as a whole. For these reasons, special purpose software may be employed by the CFD package to analyse the CAD surfaces and correct such errors to ensure a well-defined grid.

REFERENCES

1. Comstock, J.P., ed. *Principles of Naval Architecture*, Society of Naval Architects and Marine Engineers, New York, 1980.
2. Larssen, L. and Eliasson, R.E., *Principles of Yacht Design*, Adlard Coles Nautical, London, 1994.
3. Kinney, F.S., *Skene's Elements of Yacht Design*, Dodd Mead and Company, New York, 1973.
4. Faux, I.D. and Pratt, M.J., *Computational Geometry for Design and Manufacture*, Ellis Horwood Limited, Chichester.
5. Farin, G., *Curves and Surfaces for Computer Aided Geometric Design: A Practical Guide*, Academic Press, Inc., London, 1993.
6. Rogers, D.F. and Alan Adams, J. *Mathematical Elements for Computer Graphics*, McGraw-Hill Publishing Company, London, 1990.
7. Rogers, D.F., 'B-Spline Curves and Surfaces for Ship Hull Definition' in *Proceedings First International Symposium on Computer-Aided Hull Surface Definition*, The Society of Naval Architects and Marine Engineers, New York, 1977.
8. Struik, D.J., *Lectures on Classical Differential Geometry*, Addison-Wesley Publishing Company, Inc., London, 1961.
9. Pigounakis, K.G., Sapidis, N.S., Panagiotis D. Kaklis, 'Fairing Spatial B-Spline Curves', *Journal of Ship Research*, Vol. 40, No. 4, December, 1996.
10. Taylor, D.W., 'Calculations of Ships' Forms, etc.' Paper 196, *Transactions of the International Engineering Congress*, San Francisco, 1915.

11. Hoffman, D. and Zielinski, T., 'The Use of Conformal Mapping Techniques for Hull Surface Definition', in *Proceedings First International Symposium on Computer-Aided Hull Surface Definition*, The Society of Naval Architects and Marine Engineers, New York, 1977.
12. Abbott, Ira H. and von Doenhoff, A. E., *Theory of Wing Sections*, Dover Publications, Inc., New York, 1959.
13. Eppler, R., *Airfoil Design and Data*, Springer-Verlag, London, 1990.
14. Lackenby, H., 'On the Systematic Geometrical Variation of Ship Forms', *Transactions*, RINA, 1950.

CHAPTER 12
BACKGROUND TO COMPUTATIONAL FLUID DYNAMICS

S.R. Turnock[a], J. Cross-Whiter[b]
[a]Department of Ship Science and [b]Wolfson Unit for Marine Technology and Industrial Aerodynamics, University of Southampton.

12.1 INTRODUCTION

Computational fluid dynamics is the use of numerical techniques to solve the equations defining fluid flow around, within, and between bodies. The equations solved are numerical approximations to mathematical models describing the physics of the actual fluid flow. Therefore, there is always an inherent level of approximation to physical reality.

The aim of this chapter is to provide a background to the CFD techniques, which can be applied to yacht performance prediction. In this context CFD is taken to include not just the solution of the Reynolds averaged Navier-Stokes equations, but the whole range of numerical approximations available to the yacht designer for approximating the physics of the flow around realistic geometrical configurations.

The complexity of flow around a yacht explains the less rapid adoption of CFD techniques for application to marine problems than, for instance, in the aerospace industry. Often, experimental methods involving use of towing tanks or cavitation tunnels still offer a more cost-effective route. However, the advent of low-cost powerful computers has allowed progressively more complex problems to be solved numerically.

In CFD, the continuous mathematical representation of a fluid is replaced by a discrete representation. This leads to a requirement to solve equations numerically through the application of a large number of mathematical operations. The result is a solution defined at discrete positions in time and/or space. The spatial and temporal resolution of the solution is a measure of the usefulness and validity of the result. However, the cost of higher resolution is a greatly increased requirement for both data storage and computational power.

12.2 POTENTIAL FLOW METHODS

Background

The flow around yachts is principally in the high Reynolds number regime where, outside of thin boundary layers, the effects of viscosity can be neglected. If the effects of rotationality (or circulation) are confined to points, lines or sheets of singularities then the flow around bodies can be represented in terms of a fluid potential ϕ satisfying Laplace's equation everywhere within the fluid domain:

$$\nabla^2 \phi = 0 \qquad [12.1]$$

Potential flow fields can be constructed by the linear superposition of individual solutions to Equation [12.1]. Three basic singularity elements are used to represent the disturbance to the flow due to a body (see Figure 12.1). These are:

- Source element
  - where flow travels radially outwards at equal strength in all directions. A sink represents flow into a point.
- Dipole element
  - a combination of a source and a sink an infinitesimally small distance apart, which generates a solenoid flow field
- Vortex element
  - a rotational flow about a point or a line

The majority of potential flow solution methods use various combinations of these basic elements with the strength of each element determined by appropriate boundary conditions.

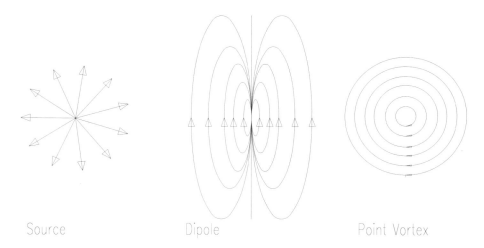

Figure 12.1 *Two-dimensional representation of potential singularities.*

Lifting-line

The flow round an aerofoil producing lift is faster on the suction (upper) surface and slower on the pressure (lower) surface. This difference in flow speed between the two surfaces can be represented by a combination of a free stream and a vortex flow circulating round the aerofoil. An aerofoil section with a prescribed circulation can be represented by a line vortex of strength Γ, called a lifting line, lying perpendicular to the chord of the aerofoil. The Kutta-Joukowski theorem states that the force on this vortex is given by:

$$F = \rho U \times \Gamma \qquad [12.2]$$

ρ = the density of the fluid
F = the force
U = free stream velocity
Γ = circulation vector

The force is directed perpendicular to the incoming flow field, and so provides lift but no drag force. For a two-dimensional aerofoil, or equivalently an infinitely long three-dimensional aerofoil, the potential flow drag is zero.

Real three-dimensional lifting surfaces, however, are not infinitely long, and do not have constant circulation along their span. Vorticity conservation laws require that at each point along a lifting line a trailing vortex is shed. This vortex is directed along the free stream, and its strength is given by

$$\gamma(y) = d\Gamma/dy \qquad [12.3]$$

At each point on a lifting line the induced velocity field is the summation of the velocities imposed by all the vortex segments. The net effect is a downwash at each point on the lifting line. This downwash is added to the freestream velocity, which rotates the inflow at the bound vortex, as shown in Figure 12.2.

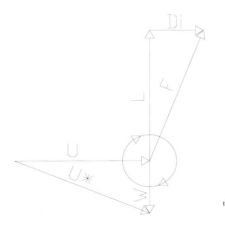

Figure 12.2 *Induced downwash and drag on lifting line.*

In Figure 12.2, W represents the induced downwash, which rotates the effective inflow from U to U*. This rotates the force on the bound vortex segment (F in Figure 12.2), which is equivalent to adding a drag vector (D_i) parallel to the freestream to the lift vector (L) perpendicular to the freestream. This is called the induced drag, since it is induced by the lift, or circulation, of the aerofoil. The total forces on a lifting line, discretised into n lifting segments, and resolved into lift perpendicular and induced drag parallel to the freestream, are then given by:

$$L = \rho U \sum_{i=1}^{n} \Gamma_i l_i$$

$$D = -\rho \sum_{i=1}^{n} \Gamma_i w_i l_i$$

[12.4]

Γ_i, w_i and l_i = the circulation, induced downwash velocity and length of the ith vortex segment

The trailing vortices are free vortices in the fluid, so cannot sustain any force. Therefore, they align with the local flow. Due to the mutual interaction of the trailing vortices they curl around each other, eventually rolling up into two strong vortices streaming aft of the wing tips. This roll-up can be calculated, leading to an iterative solution to the bound vortex plus wake flow field. However, this tends to be an unstable process, and the forces on the lifting lines are not sensitive to the exact geometry of the wake, provided the span is correct. Therefore, the wake is generally modelled as a sheet trailing downstream from the trailing edge of the aerofoil.

Vortex lattice

For computing the span-wise force distribution and the velocity field at points far from the aerofoil, given a known distribution of circulation, the lifting line approach works well. However, to compute the chord-wise force and velocity distributions, or to compute the vortex strengths from a known surface geometry, the lifting line model is extended across the aerofoil surface. At each span-wise position the chord-wise distribution of circulation is represented by a row of horseshoe vortices, like those of the lifting line described previously. The sum of these vortex strengths equals the lifting line circulation about the aerofoil at that span-wise point. Likewise, at the trailing edge the trailing vortex strengths are the summation of the trailing vortex strengths in the row.

Thus, the lifting foil is represented by a lattice of vortex segments lying perpendicular and parallel to the flow direction. Figure 12.3 shows an example of keel and rudder vortex lattice grids, including the trailing wake segments.

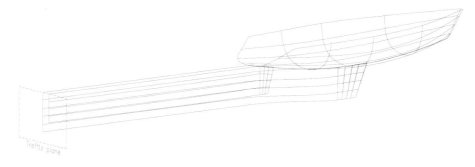

Figure 12.3 *Vortex lattice grid and wake.*

In the interior of each panel of the lattice a control (collocation) point is placed. At each control point the free stream and vortex system induce a velocity, which can be resolved into components normal and tangential to the lifting foil surface at that point. The boundary condition at each control point is that the normal velocity is zero. The velocity induced by each vortex segment is a function of its position relative to the control point, which is known, and its strength, which is a priori unknown. Thus, at each control point i the normal velocity can be represented by the summation

$$Vn_i = \mathbf{U} \cdot \mathbf{n}_i + \sum_{j=1}^{N} \frac{\Gamma_j}{4\pi} A_{ij} = 0 \qquad [12.5]$$

Vn_i = the normal velocity at control point i
$\mathbf{U}$ = the free-stream velocity vector
$\mathbf{n}_i$ = the normal vector at the control point i
A_{ij} = the influence coeficient of the jth vortex segment on the ith control point, which is a function of location

If there are N control points and N horseshoe vortices, Equation [12.5] provides a linear system of equations to solve for the unknown Γ's.

Surface panel methods

A detailed description of the surface panel method and a review of its historical development is given by Hess[1] and a comprehensive text on its application to low-speed aerodynamics is given by Katz and Plotkin.[2] The approach of the surface panel formulation is to represent the actual body surface in terms of a distribution of source and/or dipole elements on the surface with strengths appropriate to satisfy the boundary condition of zero normal flow. This allows, in comparison to the Vortex Lattice Methods (VLM), a better representation of the effects of section thickness and circulation.

Laplace's equation can be written as an integral, over the bounding surface S of a source distribution (σ) and a normal dipole distribution (μ). If **v** represents the disturbance velocity field due to the bounding surface (or body) and is defined as the difference between the local velocity at a point and that due to the free-stream velocity, then:

$$\mathbf{v} = \nabla \phi \qquad [12.6]$$

ϕ = the disturbance potential. This can be expressed in terms of a surface integral as:

$$\phi = \frac{-1}{4\pi} \iint_{S_B} \left[\frac{1}{r}\sigma - \frac{\partial}{\partial n}\left(\frac{1}{r}\right)\mu \right] dS + \frac{1}{4\pi} \iint_{S_W} \frac{\partial}{\partial n}\left(\frac{1}{r}\right) \mu dS \qquad [12.7]$$

S_B = the surface of the body
S_W = a trailing wake sheet

In equation [12.7] r is the distance from the point for which the potential is being determined to the integration point on the surface and $\partial/\partial n$ is a partial derivative in the direction normal to the local surface. A dipole distribution is used to represent the wake sheet. This can be directly related to the vorticity distribution used in VLM. The use of source and dipole distributions ensures that the velocity potential satisfies Laplace's equation everywhere outside of the body and wake and that the disturbance potential due to the body vanishes at infinity. Three boundary conditions are satisfied by ensuring that the normal component of velocity is zero on the body surface, the Kutta condition of a finite velocity at the body trailing edge is satisfied and that the trailing wake sheet is a stream surface with equal pressure either side.

The Kutta condition only applies at the trailing edge, and some other relationship has to be used to uniquely determine the distribution of μ and σ over the body. The numerical resolution of this non-uniqueness is referred to as the singularity mix of the lifting-surface method, and a number of possible schemes are in use. These are formulated in terms either of a total or disturbance value of their velocity influence or potential at a point. The perturbation potential method of Morino[3] will be described.

The numerical procedure is based on representing the body surface by a series of N quadrilateral panels, each with an unknown but constant dipole strength per unit area. The vertices of these panels are located on the actual surface of the body. The wake sheet is represented by M panels placed on the stream-surface from the trailing edge of the body surface. Its dipole strength per unit area is related to the difference in dipole potential at the trailing edge.

That is:

$$\mu_\omega = \phi_v - \phi_l \qquad [12.8]$$

ϕ_u and ϕ_l = upper and lower surface trailing edge panel potentials.

On the body surface the source strength per unit area is prescribed by satisfying the condition for zero normal velocity at the panel centroid:

$$\sigma_s = -\mathbf{U}.\mathbf{n} \qquad [12.9]$$

n = the unit normal outward from the panel surface
U = the specified inflow velocity at the panel centroid

The numerical discretisation of Equation [12.7] gives the potential at the centroid of panel i as:

$$\phi_i = \sum_{j=1}^{N}((U \cdot n_j)S_{ij} - \phi_j D_{ij}) + \sum_{k=1}^{M}\Delta\phi_k W_{ik} \qquad [12.10]$$

where for node i: S_{ij} is the influence coefficient of source panel j on node i; D_{ij} is the influence coefficient of dipole panel j on node i; W_{ik} is the influence coefficient of wake strip k extending to infinity.

As there are N independent equations corresponding to the N body surface panel centroids, Equation [12.10] is closed and can be evaluated. Expressed in matrix form it becomes:

$$[D_{ij}]\phi + [W_{ik}]\Delta\phi = [S_{ij}](U \cdot n) \qquad [12.11]$$

For Morino's Kutta condition the matrix expression [12.11] can be solved to give the vector of dipole potentials φ. Improved results are obtained using an iterative process whereby the explicit difference in pressure at the trailing edge is equated to zero. Numerical differentiation of the dipole potential along the body surface allows the surface velocities and hence pressures on the surface to be evaluated.

Again, as for the VLM, the specification of the wake shape is often kept fixed to reduce calculation time. Wake adaptation is possible but involves a significant increase in computational effort, and some form of shape control is required in order to minimise difficulties with wake roll-up generating numerical instabilities.

The time varying movement of control surfaces or the response of a vessel to a sea state can generate a wide range of fluid flow behaviours, a limited subset of which at the current time are amenable to solution. An approach such as vortex lattice or surface panel representation can be modified to include time dependent effects, as long the time variation of the flow is sufficiently slow that the main features of the flow remain similar, and it remains attached or separates only at well defined locations,

The main source of difficulty is the ability to represent the time varying development of the wake of a lifting surface. The wake strength now varies both in the span-wise and in the chord-wise (temporal) direction. Usually wake relaxation is confined to a region just downstream of the trailing edge, and at greater distances the allowed shape is prescribed in order to confine both the time for calculating the wake influence and to prevent numerical difficulties with large distortions in the surface of the wake.

Free-surface representation

The accurate calculation of the free surface position on the yacht hull and the subsequent trailing systems of diverging and transverse wave systems is essential to the evaluation of wave resistance. The free surface requires that appropriate boundary conditions are satisfied on a surface, the location of which is not known *a priori*. In addition, the wave-induced flow, when compared to zero Froude number flow, causes the sinkage and trim of the hull to alter. An exact solution requires these hull quantities to vary as the solution proceeds to ensure that the hull forces and moments are in balance.

A detailed description of the free surface boundary conditions is given in Newman.[4] On the free surface a kinematic and a dynamic boundary condition need to be satisfied. The kinematic boundary condition requires that the normal component of the fluid velocity is equal to the normal velocity of the surface itself and is identical to a solid surface condition. That is, no fluid passes through the boundary surface.

The free surface elevation at any point on the free surface can be described as a function $z = \eta(x,y,z,t)$. The kinematic boundary condition can be derived by requiring the total derivative of $(z - \eta)$ to vanish on the free surface. The result of this condition, on $z = \eta$,

$$0 = \frac{D}{Dt}(z - \eta) = \frac{\delta\phi}{\delta z} - \frac{\delta\eta}{\delta t} - \frac{\delta\phi}{\delta x}\frac{\delta\eta}{\delta x} - \frac{\delta\phi}{\delta y}\frac{\delta\eta}{\delta y} \qquad [12.12]$$

The velocity is not known on the free surface until the further dynamic boundary condition is satisfied. This condition ensures the forces acting on the air-water interface are in balance. The condition can be found from Bernoulli's equation and can be expressed in terms of the surface elevation as

$$\eta = -\frac{1}{g}\left(\frac{\delta\phi}{\delta t} + \frac{1}{2}\nabla\phi.\nabla\phi\right) \qquad [12.13]$$

A similar hierarchy of methods to that described for potential flow have been developed for free surface prediction. At its simplest, for slender bodies, the longitudinal distribution of volume in the hull is represented as Neuman-Kelvin sources distributed along the hull centre-line. The free-surface boundary condition is satisfied directly within the numerical evaluation of the influence coefficient for each source. This approach can give reasonable results but does not include detailed hull shape effects or the influence of circulation on wave generation.

The next level of improved approximation is to couple a surface panel model of the actual hull surface with Neuman-Kelvin sources mapped onto the surface the strength of which are chosen to satisfy the free-surface condition. The numerical evaluation and solution of this problem is difficult and time-consuming to formulate in a robust manner. The alternative approach, which has been increasingly adopted, is to model both the hull surface and free surface with Rankine sources and dipoles, and to panel the actual free surface or the $z = 0$ plane with appropriately modified boundary conditions. Rankine sources and dipoles are a basic form which do not contain wavemaking terms in their definitions, hence the need to panel the sea surface aswell as the hull.

A series of linear approximations are applied to simplify the solution procedure. Up to three levels of linearisation can be applied. The first of these simplifies the kinematic condition by neglecting the changes in surface slope in the x and y direction and then equates only the vertical velocities of the free surface and fluid particles. The dynamic boundary condition can be linearised by neglecting the second order term in Equation [12.13]. A further assumption is to apply the boundary condition, not on the actual surface $z = \eta$, but on the $z = 0$ plane, with a suitable modification in terms of a Taylor series expansion which assumes η is small.

Boundary layer approximations

Viscous effects can be included in potential calculations through iterative coupling of a boundary layer prediction to a potential flow solution. The momentum integral equation for the boundary layer is used with appropriate assumptions for the velocity profile based on the solution of the outer potential flow. Different velocity profiles are required for the laminar and turbulent regions. Difficulties with predicting the location of transition from laminar to turbulent flow and over what distance this transition occurs require the use of appropriate, empirically derived expressions. These calculations are usually carried out along potential flow streamlines traced on the surface of the actual three-dimensional geometry.[5] Such an approach gives a good prediction of the skin friction, provided cross-flow terms do not have a strong influence on the boundary layer. Knowledge of how the boundary layer is developing can be used to identify the likely occurrence of separation.

Two approaches are possible for coupling the boundary layer development to the potential flow field. In the first, the surface panels are located on the edge of the boundary layer region defined by the displacement thickness δ^*. The same rationale can be used to represent regions of well defined separation. For instance, on a mast-sail geometry the flow on the low pressure surface is likely to separate at some position just behind the point of maximum thickness. This free shear layer will reattach at some point on the sail surface. Flow within the zone of separation is likely to be slow moving, and the pressure can be considered as close to the value at the point of separation. If panels are located on the estimated location of the free shear layer a good potential flow solution can be obtained. The second coupling approach is based on representing the effect of the boundary layer as a change in momentum of the fluid flow due to viscous shear. This is equivalent to a surface transpiration velocity. The magnitude of this transpiration velocity V_n can be found directly from the definition of displacement thickness δ^* as follows:

$$V_n = \frac{-\delta(U_e \delta^*)}{\delta s} \qquad [12.14]$$

U_e = the tangential velocity at the panel centroid
s = the direction along a streamline

An iterative process is used. The potential flow is first solved. An appropriate boundary layer model is used to calculate δ^* and the new surface panel position or the required transpiration velocity. This is then imposed as a change to the surface boundary condition and the process repeated until convergence of the overall flow is achieved. Using these methods a reasonable prediction of both skin friction and pressure form drag can be calculated, provided the flow regime avoids regions of separation and that the flow is predominantly two-dimensional along stream lines.

12.3 NAVIER-STOKES

Reynolds averaged Navier-Stokes (RANS) equations

The behaviour of a fluid is governed by the need to conserve mass, momentum and energy. Additionally, relationships have to be established between the physical properties of the fluid medium (density, viscosity). The physical properties and laws of fluid motion can be used to derive a mathematical formulation which allows viscous and rotational flow features to be included.[6,7,8]

Continuity equation

Conservation of mass applied to fluid passing through an infinitesimal fixed volume leads to

$$\frac{\delta \rho}{\delta t} + \nabla.(\rho V) = 0 \qquad [12.15]$$

For an incompressible fluid (water) this reduces to

$$\nabla.V = 0 \qquad [12.16]$$

in cartesian coordinates $\nabla = (\delta/\delta x, \delta/\delta y, \delta/\delta z)$

Air can be considered incompressible for V < 100 m/s.

Momentum equation

Again, applied to an infinitesimal fixed volume the Navier-Stokes equations for the conservation of momentum can be reduced for an incompressible Newtonian fluid (fluid is isotropic and stress at a point is linearly dependent on the rates of strain of the fluid) with constant viscosity to:

$$\rho \frac{DV}{Dt} = \rho \left(\frac{\delta V}{\delta t} + (\nabla.V)V \right) = \rho f - \nabla p + \mu \nabla^2 V \qquad [12.17]$$

where the *left-hand side* of [12.17] is the substantial derivative of the velocity; on the *right-hand side* the first term is the body force f acting on the fluid, usually only gravity g for water, the second term is the local pressure gradient, and the third term is derived from the local surface viscous stress.

Direct solution of the above equations is starting to be possible for very specific applications and at small Reynolds numbers (~500). However, when applied to unsteady turbulent flow, the turbulent eddy motions that need to be resolved are at such a fine scale that direct solution of practical turbulent flows will not be possible for the foreseeable future.

One alternative approach is to use a coarser grid to resolve the larger scale of eddies and a sub-domain model to represent the time dependent effects of the sub-grid scale turbulence. This Large Eddy Simulation (LES) approach, although at an early stage of development, has been successfully applied to ship flows by Miyata.[9]

The most common approach to the solution of the Navier-Stokes equations is based on a time-averaged approach. The unsteady flow, due to the presence of turbulence, is modelled by resolving into time mean and fluctuating components and then time averaging the whole system. The assumption is made that the new terms introduced in these Reynolds averaged Navier-Stokes (RANS) equations can be related to apparent stress gradients associated with turbulent motion. The new quantities have to be related to the mean flow variables through turbulence models. It is this assumption which introduces most of the uncertainties in RANS flow solutions.

Flow variables are represented by their time-averaged value plus a fluctuating component about the average, e.g.

$$q = \bar{q} + q' \qquad [12.18]$$

Fluctuations are assumed to be random. Time-averaging eliminates quantities which involve only one fluctuating component but not those involving two, i.e.

$$\overline{q'} = 0, \quad \overline{q'q'} \neq 0 \qquad [12.19]$$

The conservation of mass equation is identical to [12.16] but with the original flow variables replaced by their time averaged values. The form of the Reynolds averaged momentum equation is more complex and will be illustrated with respect to its form in compact tensor notation:

$$\frac{\delta}{\delta t}(\overline{\rho u_i}) + \frac{\delta}{\delta x_j}(\overline{\rho u_i u_j}) = -\frac{\delta \bar{p}}{\delta x_i} + \frac{\delta}{\delta x_j}\left(\overline{\tau_{ij}} - \overline{\rho u'_i u'_j}\right) \qquad [12.20]$$

where the laminar stress tensor τ_{ij} is

$$\overline{\tau_{ij}} = \mu\left(\frac{\overline{\delta u_i}}{\delta x_j} + \frac{\overline{\delta u_j}}{\delta x_i}\right) \qquad [12.21]$$

The apparent stresses due to the turbulence are referred to as the Reynolds stresses and can be written in terms of a turbulent stress tensor

$$\left(\overline{\tau_{ij}}\right)_{turb} = -\rho\overline{u'_i u'_j} \qquad [12.22]$$

which represents the apparent stresses due to the transport of momentum by turbulent fluctuations and deformations attributed to those fluctuations. The Reynolds equations cannot be solved unless the new turbulence-induced stresses can be related to the time-mean flow variables. This problem of closure is usually achieved by the turbulence model.

Turbulence modelling

The suitability of a chosen turbulence model and the detail and accuracy with which it represents turbulent flow behaviour are crucial to the success of a RANS flow solution. There is no universally applicable turbulence model or even agreed approach. The various forms of model will be briefly described. With close attention the results of a calculation can be made to agree with experimentally observed behaviour by manipulating the form and empirical constants of a turbulence model. However, such an approach does not guarantee that a similarly close comparison will be obtained with even a slightly different flow geometry. It is also difficult to determine whether the poor behaviour of a given turbulence model is due to lack of grid resolution or the assumptions of the model.

A systematic approach is the best methodology to apply. Knowledge of the behaviour of different turbulence models for different flow configurations requires extensive experience. Examining the sensitivity of the solution to the choice of model and of model parameters is computationally costly but allows a greater confidence level to be attached to the results. New users of RANS flow solvers should attempt to develop an understanding of how critical the turbulence model is to their particular analysis.

A range of turbulence models are available with commercial flow solvers, usually with access which allows users to supply alternate values for the various constants. Models can be classified into three different levels: ones which use the Boussinesq assumption or turbulent viscosity models, ones which attempt to close the Reynolds equations, known as Reynolds stress or stress-equation models, and finally those based on Large Eddy Simulation. The higher the classification the greater the computational effort required. Most engineering applications still use the first category.

Boussinesq assumed that the apparent shearing stresses can be related to a rate of mean strain through a scalar turbulent or 'eddy' viscosity. The Reynolds stress tensor can then be expressed as

$$-\rho\overline{u'_i u'_j} = \mu_T \left(\frac{\delta u_i}{\delta x_j} + \frac{\delta u_j}{\delta x_i} \right) - \frac{2}{3}\delta_{ij}\left(\mu_T \frac{\delta u_k}{\delta x_k} + \rho\overline{k} \right) \qquad [12.23]$$

μ_T = the turbulent viscosity
k = the kinetic energy of turbulence
δ_{ij} = Kronecter delta (i = j, δ_{ij} = 1, i ≠ j, δ_{ij} = 0)

$$\bar{k} = \frac{\overline{u'_i u'_j}}{2} \quad\quad [12.24]$$

This first, Boussinesq, category of models can be further sub-divided by the number of additional partial differential equations which need to be solved in order to close the model with regard to the value of μ_T and k. These range from zero-equation (algebraic) models such as the Baldwin-Lomax mixing length model through to two equation models such as the widely used k–ε approach.

The type of model chosen will depend on a number of factors, including the local Reynolds number, surface roughness, pressure gradient and whether it is shear layer flow away from the body surface. Many models use two different approximations for an inner and outer region of the boundary layer. Such 'law of the wall' models require a specified range of thickness for the first layer of computational cells next to the wall, which is determined by a non-dimensional boundary layer thickness parameter y+. This requirement needs to be used as part of the grid generation process in order to ensure that the turbulence model is applied correctly.

12.4 SOLUTION TECHNIQUES

Analytical solution is only possible for a limited range of potential flows and even fewer viscous flows. Solution of the flow at all locations within a domain is replaced by the solution at a discrete number of nodes. The partial differential equations defining the flow are replaced by difference equations based on the spatial separation of the solution nodes. The definition of these solution nodes is referred to as the grid generation process. The finer the node spacing the lower the discretisation error of representing a continuous function at discrete intervals, but the larger the number of unknown values which need to be found.

The two main approaches to solving a fluid dynamics problem require either the definition of boundary elements on the surface defining the flow domain or nodes distributed within the flow domain. The complexity of a field method is proportional to the number of spatial nodes (N^3 for three-dimensional), and is generally at least an order of magnitude more difficult than the corresponding boundary element approach (N^2 for three-dimensional).

The potential methods described in Section 12.2 are all boundary element methods. Their solution requires the calculation of influence coefficients among the various panels and panel centroids. The calculation of these influence coefficients can be mathematically intensive, especially for Neumann-Kelvin free surface sources. The resulting, usually linear, system of equations is evaluated using standard methods. These are either direct solution via matrix inversion or an

248 *Sailing Yacht Design: Theory*

iterative process of estimating the solution and progressively correcting to obtain the actual solution to within a defined limit of convergence.

The numerical solution of the incompressible Reynolds averaged Navier-Stokes equations can be carried out using a variety of techniques. The correct construction of the numerical difference equations is critical to the accuracy of the solution obtained. In general, the type of method will be related to the type of flow domain and the way it is discretised. Standard approaches are:

- *Finite difference*
  - where the equations are constructed for the solution at a point in space
- *Finite volume*
  - where an average value of flow properties is found in a discrete volume surrounding the control node
- *Finite element*
  - where simple functions are used to represent the variation in flow properties between nodes

Each approach has various advantages and disadvantages although the finite volume/element methods are the most widely applied to real engineering applications.

Whichever method is used, the non-linear, coupled nature of the incompressible Navier-Stokes equations results in the need to pre-condition the equations in order to be able to obtain a solution. The main approach used is to iteratively solve for each flow variable with the others held constant and then to update all the variables via an outer iterative loop. A number of variations are in use (SIMPLE, SIMPLER, PISO) but the non-linear nature of the coupling between pressure and velocity results in a large number of iterative cycles being necessary in order to obtain a converged solution. The quality of the grid discretisation affects the speed of solution, and poor grids require larger under-relaxation parameters, which reduce the correction applied at each iteration so that grid induced errors do not grow and cause the solution to diverge. Another approach, which has some advantages for unsteady problems, is to introduce an artificial compressibility in order to facilitate solution.

12.5 GRID GENERATION

In order to analyse a particular fluid dynamic or structural problem it is necessary to be able to define the geometry of the problem. The format of this geometry definition will usually be specific to a particular numerical computer code. It is important to appreciate that the quality of the geometry definition will be critical to the accuracy of the numerical solution obtained. The generation of the problem definition usually consumes the majority of user time when practical problems are solved using numerical analysis methods.

The alternative approaches to grid generation[10,11] are structured, in which the available space/surface is divided into regular intervals or unstructured, where no regularity is imposed. The structured approach reduces the storage space required to map the physical domain onto the computational domain, and can improve the accuracy and ease of solution of the numerical method. However, it is often difficult to impose regularity on all possible topologies, especially in three-dimensions.

All methods require connectivity, whether it is implicit (structured) or explicit (unstructured) between spatial nodes (2D or 3D), lines, surfaces and volumes (3D). The grid generation process requires a defined spatial domain, which contains the body of interest and an appropriate outer limit to the region in which the flow equations are to be solved. This spatial solution domain should then be subdivided into connected nodes, lines, surfaces (and volumes) in such a way that the influence of the grid on the eventual flow solution is minimised. For a complex three-dimensional geometry using a structured approach it is often essential that the initial domain is further sub-divided prior to the detailed grid generation. The generation of these sub-domains, often referred to as a multi-block approach can be time consuming, and it is only recently that semi-automatic methods have been developed and applied to specific vessel types.

It is worth noting that a multi-block method can also be used to couple structured and unstructured grid generation techniques in a hybrid approach. Such an approach can be used to ensure that the most appropriate grid is used for a given local flow regime. A further refinement is to use the evolution of the flow solution to control definition of the local grid. Referred to as grid adaptation this requires the coupling of the flow solver to the underlying body geometry.

Structured grid generation

The base elements used are a quadrilateral panel (surface) and hexahedral cell (volume). These elements naturally map onto two- or three-dimensional regular computational memory arrays of node values, with the connectivity defined by the node's position within the array.

The regularity of the structure imposes 4 boundary edges for a surface and 6 bounding faces for a volume. Grid generation requires the sub-division process be carried out in the following order:

1. Each boundary edge is sub-divided into the required divisions. This can be achieved either in equally spaced divisions or by imposing a stretching function.
2. The sub-division of the boundary edges is used to generate the interior face (surface) nodes. An interpolation function is used to generate the nodal values. The grid can be improved using techniques such as elliptic grid refinement, in which a system of partial differential equations are solved.
3. In three-dimensions all six faces are subdivided into quadrilateral panels and the same techniques are used to fill and refine the interior nodes of the domain.

Once the grid has been generated further refinement has often to be imposed. For instance, when using law of the wall turbulence models it is essential that the first cell out from a solid boundary lies within a given distance defined by y+. Final inspection, usually visual with the aid of grid quality analysis software, will indicate areas of poor grid quality and where the grid generation process has to be repeated with appropriate modification.

Unstructured grid generation

The base element surface is a triangle and corresponding volume a tetrahedron. In many unstructured grid generation processes only elements of these types are used, as any other shape can be derived using them. The two main approaches are that of the 'advancing front' and 'Delaunay triangulation'. The advancing front technique grows tetrahedral cells from a face into the interior domain or triangles into the interior domain from the boundary edges. The Delaunay algorithm recursively divides a region by ensuring that the nodes placed for the next level generate elements which are as nearly equilateral as possible.

12.6 VISUALISATION AND VALIDATION

The purpose of CFD analysis is to derive information of use to the designer. This can either be qualitative in the form of flow visualisation or quantitative values of forces and surface pressures.

Methods for data presentation and visualisation

Information derived from forces and moments can be treated in the same way as that obtained from traditional experimental and analytical approaches. Graphical presentation, in terms of appropriate non-dimensional quantities, is recommended for gaining physical insight into performance.

The ability of all but the simplest CFD technique to solve large numbers of flow variables at many spatial and temporal locations presents the designer with a previously unobtainable knowledge of local flow behaviour. Fortunately, like CFD itself, visualisation of data using computational analysis techniques has also benefited from advances in computer technology.

The general types of visualisation techniques are as follows:

- Streamlines or stream surfaces
- Velocity vectors
- Contour plots of scalar quantities on body surface or user defined planes
- Particle Tracking

These techniques mainly provide qualitative information. Use of the software to capture flow features of interest and the ability to compare behaviour before and after a design change can still be a difficult process.

Methods for determining forces

The solution of the flow equations can be used to find the pressure, velocity and shear stress on body surfaces. Integration of these quantities over the complete body surface allows the total force and moments to be obtained. For instance for a body of N surface panels the total force can be found as

$$F_P = \sum_{i=1}^{N} \frac{1}{2}\rho V_\infty^2 A_i n_i Cp_i + \sum_{i=1}^{N} \frac{1}{2}\rho V_\infty^2 A_i v_i Cf_i$$

$$M_P = \sum_{i=1}^{N} \frac{1}{2}\rho V_\infty^2 A_i (r \times n_i) Cp_i + \sum_{i=1}^{N} \frac{1}{2}\rho V_\infty^2 A_i (r \times v_i) Cf_i$$

[12.25]

for panel i
- A_i = the area
- n_i = the unit normal to the panel
- r_i = the position vector of the panel centroid
- v_i = the unit vector in the direction of the local flow
- C_{fi} = the local skin friction coefficient
- C_{pi} = the local surface pressure coefficient

Inaccuracies in this method arise in regions where the surface curvature is rapidly changing, for instance at the leading edge of a keel, and there are insufficient panels to accurately follow the surface. The drag force is particularly sensitive to such errors, as it is effectively the difference in pressure acting on the fore and aft half of a section, a small difference between two large quantities.

For free surface methods, if the free-surface boundary condition is satisfied fully then the contribution to resistance will be included in the integration of surface pressures and shear stress.

An alternative method is to calculate the change in flow parameters at some plane downstream of the body and thereby infer the momentum change in the fluid due to the body and hence calculate the force acting on the body. Such an approach is referred to as a Trefftz plane analysis. For VLM and surface panel wakes the drag and lift can be calculated directly from the circulation contained in the wake. For RANS solutions an equivalent to an experimental wake traverse has to be conducted, with the values of total pressure integrated to find the total momentum loss. The Trefftz plane calculations provide an efficient means of computing lift and induced drag, using far fewer panels than a force integration, provided that wake contractions around the aft end of the hull, and thus the lifting line spans, are modelled correctly.

Validation of computed results

A mathematical model of a physical process generally involves a degree of approximation. In using such a model it is necessary to appreciate the confidence with which the model can be used. Analogous to the errors (or more correctly the degree of uncertainty) in the acquisition of experimental data, numerical modelling gives rise to uncertainties in the answer obtained. The process of validation attempts to quantify these uncertainties. Three stages are generally required:

1. Verification of the code implementation against the underlying mathematical formulation. This ensures the code is free of error. Ideally the comparison should be made with an analytic solution.
2. Investigation of the independence of the solution from numerical parameters: the most common form of dependence is the grid density. Normally, the number of grid points is increased until the solution does not change.
3. Comparison of numerical and experimental data: CFD codes are an approximation to the actual physics of the flow, so there will be differences between the experimental and numerical results. Experimental data should have a specified accuracy, which should then allow the differences between experiment and theory to be quantified. However, in many codes some degree of empiricism is used to adjust the numerical model to fit specific experimental data. The extent to which such an empirically adjusted model can be said to be valid for cases run at different conditions requires careful consideration.

12.7 COMMON APPLICATIONS

Sails

Currently the main method applied to realistic sail configurations is the vortex-lattice approach. Reasonable predictions of sail performance can be achieved particularly when coupled to a structural sail model. The limitations are principally the inability to include the influence of flow around the mast and its associated separation zones and the restriction to upwind conditions. Surface panel formulations can be applied to complex sail configurations, with the possibility of including boundary layer interaction, but still in the upwind configuration only.

The main restriction to the application of RANS solvers to the flow around multiple sails is the difficulty of generating grids of sufficient quality and fineness around the topology. Downwind performance prediction is restricted solely to RANS codes. However, even in these codes accurate prediction of the driving force suffers because of the still as yet poor performance of turbulence models in predicting form and viscous drag.

Figure 12.4 shows the surface pressure distribution on an IACC rig, calculated using a surface panel code including the mast but not the modelling of separation zones.

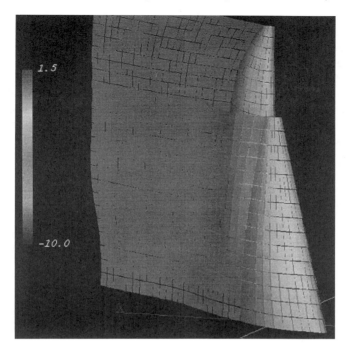

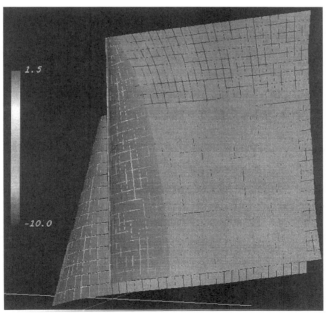

Figure 12.4 *Surface panel solution for genoa and mainsail; leeward (top), windward (bottom).*

Hulls

In the downwind condition good progress is being made with fully non-linear surface panel codes being used with reasonable accuracy to predict wave-making resistance. However, for upwind conditions the influence of yaw and hull–keel interaction may cause difficulties unless the wave-making effects of circulation can be included. The computational effort required to include free-surface and viscous effects are large and as yet only limited RANS calculations have been carried out. A typical yacht hull and appendage, surface panel geometry, is shown in Figure 12.5.

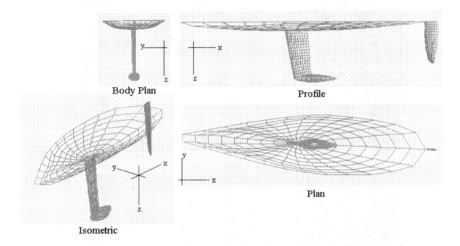

Figure 12.5 *Hull and appendage geometry for a typical yacht.*

Appendages

Appendages are the main area in which the whole range of available CFD tools can be applied. This is in general only when the effect of the free-surface interaction can be neglected. The confidence and knowledge with which similar configurations are analysed in the aerospace industry is of great benefit. Design of keel and bulb shape, performance of rudders, selection of appropriate sections can all be carried out and optimised for given performance parameters. Figures 12.6, 12.7 and 12.8 show a comparison of suface pressures on a typical rudder with a surface panel code, Euler code and commercial RANS code with linear k-E turbulence model.

Figure 12.6 *Surface panel method solution.*

Figure 12.7 *Euler solution.*

256 *Sailing Yacht Design: Theory*

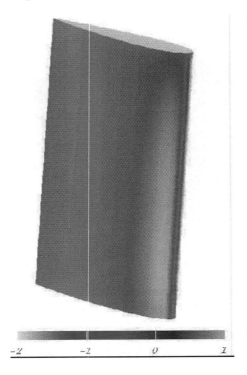

Figure 12.8 *RANS solution.*

12.8 CLOSURE

Progress in the application of CFD techniques to maritime and in particular yacht analysis has been rapid in recent years. This has been driven primarily by the rapid reduction in computational costs. Surface panel methods incorporating boundary layer approximations, free surface prediction and unsteady effects are widely available and will give good performance even on low end work stations and the latest generation of personal computer systems.

However, the cost of purchasing/using such software can be prohibitive, especially for the most sophisticated RANS flow solvers which reflects the significant amount of research and development necessary to bring such products to the market place. Additionally, the time required to be fully versed in their use and to be able to apply them with confidence as part of the yacht designers suite of analysis tools can be significant.

These costs have to be compared with the potential benefits of better designs based on informed knowledge of the actual flow. It should be possible, if the flow analysis tools are applied properly, to reduce the risk of poor performance in yachts and to more tightly focus yacht towing tank and wind tunnel testing. Computer performance per unit cost is still reducing rapidly. The use of CFD is therefore likely to increase if not predominate in the area of hydro-aerodynamic analysis of yacht performance. However, the most important factor will always remain the skill of the analyst in interpreting the results of the myriad calculations.

REFERENCES

1. Hess, J.L., *Panel Methods in Computational Fluid Dynamics*, Annual review of Fluid Mechanics, Vol. 22, 1990.
2. Katz ,J. and Plotkin, A., *Low-Speed Aerodynamics, From Wing Theory to Panel Methods*, Mc-Graw-Hill Inc., 1991.
3. Morino, L. and Kuo, C-C., *Subsonic potential aerodynamics for complex configurations: a general theory*, AIAA Journal, Vol. 12, No. 2, 1974.
4. Newman, J., *Marine Hydrodynamics*, MIT Press,
5. Larsson, L., *Scientific Methods in Yacht Design*, Annual review of Fluid Mechanics, Vol. 22, 1990.
6. Anderson, D.A., Tannehill, J.C. and Pletcher, R.H., *Computational Fluid Mechanics and Heat Transfer*, Hemisphere, 1984.
7. Anderson, J.D., *Computational Fliud Dynamics. The Basics with Applications.* McGraw-Hill, 1995.
8. Fletcher, C.A.J., *Computational Techniques for Fluid Mechanics*, Springer-Verlag, 1991.
9. Miyata, H. Sato, T. and Baba, N., *Difference Solution of a viscous flow with free-surface about an advancing ship,* Journal of Computational Physics, Vol. 72, 1987.
10. Eiseman. P.R., *A multi-surface method of coordinate generation.* Journal of Computational Physics, Vol. 33, 118–150, 1979.
11. Nakakashi, K. and Sharon, D., *Hybrid prismatic.tetrahedral grid generation for viscous flow applications.* AIAA Journal, Vol 36(2), 157–162, 1998.

CHAPTER 13
BACKGROUND TO FINITE ELEMENT ANALYSIS

R. Loscombe and A. Shenoi*
Southampton Institute and *University of Southampton

13.1 THE ROLE OF FEA IN DESIGN

Complexities in design

The structural design of an efficient sailing yacht hull involves the analysis, re-analysis and refinement of a complex structure under complex loading until a set of demanding criteria have been satisfied. This statement can be justified by examining the characteristics of yacht structures.

Complex structure

Local hull shell stiffness derives as much from curvature and irregularly arranged interior mouldings/furniture as from stringers and frames. Curvature implies the need for a non-linear analysis.

Construction materials exhibit a high degree of spatial variability due to the intentional orientation of fibres to maximise performance plus the inherent statistical variation of mechanical properties due to the use of natural materials (wood veneers) and variations in fibre volume fraction and fibre misalignment.

The hull girder generally has low length to depth and length to beam ratios making a simple beam approach to hull girder strength/stiffness a little suspect.

The use of low modulus-high strength materials allows designers to utilise membrane effects which again suggests a non-linear analysis is needed.

The ease with which complex mouldings can be made, translates into a problem for the structural designer in trying to idealise the geometry.

Complex loading

The hull is subjected to numerous concentrated load points, e.g. at the mast step, the stemhead fitting, chainplates and winches. These not only lead to high stress gradients, but also mean a simple analysis based on local pressure loads alone may be suspect.

Some loads are related to hull girder stiffness. For example, pre-tensioning of the standing rigging causes hull distortion (the so called 'banana' mode) which relieves the tension and so unloads the hull girder.

Dynamic loading can be a function of hull elasticity and mass distribution. For example the inertial load and dynamic magnification experienced by an aero-rig due to rolling and pitching will be a direct function of the mast mass and stiffness distribution.

Demanding design criteria

A high degree of hull stiffness is required to enable the rig to be highly tensioned. A high degree of lateral and inplane panel stiffness is required. This is particularly so for sandwich decks in order to avoid wrinkling/panel buckling and provide the crew with confidence as they work around the deck.

The bottom structure must be able to support the keel rigidly, yet permit grounding loads to be distributed throughout the bottom structure with minimal cracking. In addition, the keel bolts, floor and girder structure should be capable of resisting fatigue loading, e.g. due to rolling in the running condition.

Adequate shell robustness must be provided in order to resist impact with flotsam. This is a particularly delicate matter for sandwich hulls, since a minimum weight criterion drives the designer to adopt a minimum face thickness and 'core shear strength' governed configuration.

The need for complex modelling capability

The designer of larger, conventional ships is usually dealing with fairly slender hulls which approximately obey simple beam theory. The hull girder is largely composed of flat stiffened panels which are amenable to design and analysis using rule type formulations. A large experiential database is also available which allows quite simple design methods to be used with confidence.

The classification societies also provide invaluable design guides for sailing boat designers,[1] but the complexities described in the previous section limit the extent to which a simple rule constitutes a completely adequate design method. The process of design refinement (note the term 'optimisation' has been avoided here) for sailing yacht structures requires design/analytical methods capable of reflecting the effect of changes in the potentially large number of input parameters. For example, the number of parameters needed to define a fibre-reinforced plastic sandwich shell can be an order greater than that needed to specify a flat steel or alloy panel.

Not surprising then, that classification societies offer designers the option of a 'direct calculation' route to plan approval. Finite element analysis (FEA) constitutes one component of such an approach. FEA is essentially a tool for estimating the response of numerical structural models to known loading. It is probably the only method capable of analysing structures of the complexity of sailing yachts, short of instrumenting physical models. This statement should not be taken as meaning that less numerically intensive methods can be dispensed with. On the contrary, FEA requires the structure to be initially 'designed' and such methods are essential for providing 'first guess' level scantlings (see Chapter 9 on Structural Design of Hull Elements). The design process is one of iteration.

The other component of direct calculation is load prediction. Key loads are bottom and bow slamming and surfing design pressures and rig loads. These cannot be obtained by FEA. Load prediction is dealt with elsewhere in the book. Designers may use classification society rule pressure algorithms and in-house methods for rig loads as input for FEA, although classification society pressures cannot be divorced from their associated design methods.

Advances in computational technology

The origins of finite element date back to the early 1940s[2] and some might say earlier if the basic theorems are included. FEA is routinely used in aero-structures design offices, less so in ship design offices and perhaps only on an occasional basis in yacht design work. The reasons behind this are not hard to find. Although yacht and airframes are both thin skinned, stiffened shell structures made of a wide range of materials, capital costs are much lower for the former resulting in a smaller design budget, making FEA based design more difficult.

The key constraints on implementing FEA into the structural design of sailing yachts are:

- hardware requirements
- modelling and interpretation man-hours

Personal computers of quite modest specification are capable of providing worthwhile results.[3] Many commercial packages utilise a database system and allow import of CAD files which significantly reduces modelling times. In common with lines development packages, interpretation, thankfully remains in the hands of the designer.

Other advances in computational technology have been concerned with improving the efficiency of the numerical procedures which underpin a FEA. It is not appropriate to go into these in detail in this short introduction to FEA. However, given the need for designers to interpret what a FEA is telling them, it is essential that any package is not treated as a 'black box'. Unlike with drafting packages, 'what to see' is seldom 'what you get' in a finite element analysis. The next section therefore aims to provide an overview of FEA and the final section discusses some aspects of modelling.

13.2 FEA THEORY – THE STIFFNESS METHOD

The spring and bar analogy

There are thousands of papers and books on the subject of FEA. FEA is an all embracing term and covers a wide range of physical problems. It is possible to analyse heat transfer, electro-magnetic behaviour and fluid mechanics.[4]

However, in the context of sailing yacht structures, it is probably not unreasonable to say that the overwhelming majority of analyses are linear static structural ones with a lesser number of non-linear analyses. This will be the focus of this chapter.

Consider a bar or truss element loaded and free to displace axially.[5]

Figure 13.1 *Two noded spring element.*

   L = element length
   A = cross-sectional area
   E = Young's Modulus = stress (σ)/strain (ε)

Stiffness (k) is the load required to cause a unit displacement, where $k = p/u = \sigma A/\varepsilon L = EA/L$. The net displacement of the element at **nodes** 1 and 2 is $u_1 - u_2$ hence $p_1 = k(u_1 - u_2)$ and by equilibrium $p_1 = -p_2$. The relationship between force and displacement for a two-dimensional truss element is given by;

$$\begin{Bmatrix} p_1 \\ p_2 \end{Bmatrix} = \frac{EA}{L} \begin{bmatrix} 1 & -1 \\ -1 & 1 \end{bmatrix} \begin{Bmatrix} u_1 \\ u_2 \end{Bmatrix} \equiv \{p\} = [k]\{u\} \qquad [13.1]$$

Notice how the forces and displacements are coaxial; they are referenced to the **element coordinate system** where local x corresponds to the element axis. Now consider a framework made up of three such truss elements, Figure 13.2.

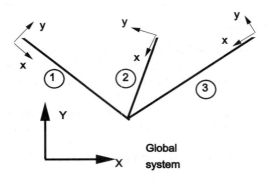

Figure 13.2 *Three truss element framework.*

It is necessary to work in the **global coordinate system**. This means that the local stiffness must be transformed from the local system to the global system. Consider the nodal displacement shown in Figure 13.3. It follows from this diagram that displacements in the local system (u,v) can be written in terms of displacements in the global system (U,V), i.e.

u = U cos θ + V sin θ [13.2]

v = V cos θ − U sin θ

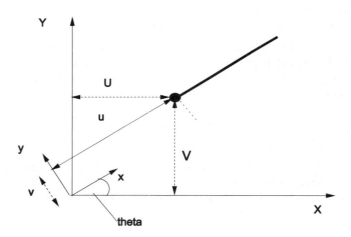

Figure 13.3 *Node under loacl and global displacements.*

In matrix form, this would be:

$$\begin{Bmatrix} u \\ v \end{Bmatrix} = \begin{bmatrix} \cos\theta & \sin\theta \\ -\sin\theta & \cos\theta \end{bmatrix} \begin{Bmatrix} U \\ V \end{Bmatrix} \quad [13.3]$$

or $\{u\} = [\xi]\{U\}$ where $[\xi]$ denotes the **transformation** matrix. Since a truss element has two nodal points, the relationship between global and local displacement is:

$$\begin{Bmatrix} u_1 \\ v_1 \\ u_2 \\ v_2 \end{Bmatrix} = \begin{bmatrix} \xi & 0 \\ 0 & \xi \end{bmatrix} \begin{Bmatrix} U_1 \\ V_1 \\ U_2 \\ V_2 \end{Bmatrix} \quad [13.4]$$

A similar approach can be used for transforming the force vector,

$$\{p\} = \begin{bmatrix} \xi & 0 \\ 0 & \xi \end{bmatrix} \{P\} \quad [13.5]$$

Hence if $\{p\} = [k]\{u\}$ refers to the local system and $\{P\} = [K]\{U\}$ refers to the global system. Then,

$$[K] = \begin{bmatrix} \xi & 0 \\ 0 & \xi \end{bmatrix}^T [k] \begin{bmatrix} \xi & 0 \\ 0 & \xi \end{bmatrix} \quad [13.6]$$

where T denotes the transpose and [K] represents the transformed stiffness matrix. Note: $[\xi]$ is orthogonal and [K] is now a 4×4 matrix.

The transformed element stiffness matrix simplifies to:

$$[K] = k \begin{bmatrix} c^2 & cs & -c^2 & -cs \\ cs & s^2 & -cs & -s^2 \\ -c^2 & -cs & c^2 & cs \\ -cs & -s^2 & cs & s^2 \end{bmatrix} \quad [13.7]$$

$c = \cos\theta$
$s = \sin\theta$
$k = EA/L$

The three truss problem of Figure 13.2 now proceeds by transforming each element stiffness into global terms and summing terms at each degree of freedom. Figure 13.4 shows how each element contributes to the overall stiffness matrix. Cells which do not contain such a contribution are null.

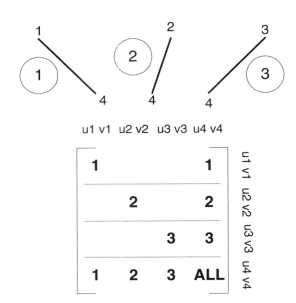

Figure 13.4 *Structural stiffness matrix by superimposition of element stiffness matrices.*

In general, a structural stiffness matrix will be highly banded. The problem size is controlled as much by the half bandwidth as by the total number of degrees of freedom. FEA programmers have devoted much effort to developing algorithms which optimise the ordering of elements. This process is known as **re-ordering**. The three frame structure is constrained at nodes 1, 2 and 3 and hence only u_4 and v_4 are non zero.

In attempting to solve for the unknown displacements it is essential to use the **constrained structural stiffness matrix**.

$$\begin{Bmatrix} P \\ R \end{Bmatrix} = \begin{bmatrix} K_{11} & K_{12} \\ K_{21} & K_{22} \end{bmatrix} \begin{Bmatrix} U \\ 0 \end{Bmatrix} \quad [13.8]$$

The displacement {U} and reaction {R} vectors can be obtained from:

$$\{U\} = [K_{11}]^{-1}\{P\} \quad [13.9]$$

$$\{R\} = [K_{21}]\{U\}$$

The calculation steps for a typical linear finite element analysis are shown in Figure 13.5.

Background to Finite Element Analysis

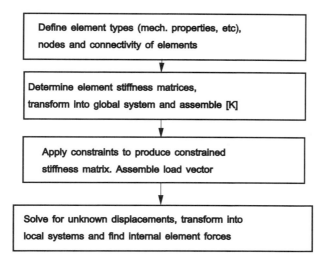

Figure 13.5 *Finite element method flow chart.*

Stiffness matrix for a three-dimensional beam element

Strength of materials theory may be used to develop a stiffness matrix for a three-dimensional beam which will of course have twelve degrees of freedom.

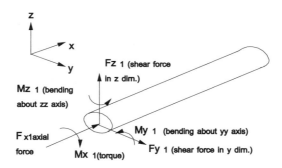

Figure 13.6 *Three-dimensional beam element 'forces'.*

The stiffness matrix looks like,[5]

$$\begin{Bmatrix} F_{x1} \\ F_{y1} \\ F_{z1} \\ M_{x1} \\ M_{y1} \\ M_{z1} \\ F_{x2} \\ F_{y2} \\ F_{z2} \\ M_{x2} \\ M_{y2} \\ M_{z2} \end{Bmatrix} = \begin{bmatrix} \frac{EA}{L} & 0 & 0 & 0 & 0 & 0 & -\frac{EA}{L} & 0 & 0 & 0 & 0 & 0 \\ 0 & \frac{12EI_z}{L^3} & 0 & 0 & 0 & \frac{6EI_z}{L^2} & 0 & -\frac{12EI_z}{L^3} & 0 & 0 & 0 & \frac{6EI_z}{L^2} \\ 0 & 0 & \frac{12EI_y}{L^3} & 0 & -\frac{6EI_y}{L^2} & 0 & 0 & 0 & -\frac{12EI_y}{L^3} & 0 & \frac{6EI_y}{L^2} & 0 \\ 0 & 0 & 0 & \frac{GJ}{L} & 0 & 0 & 0 & 0 & 0 & -\frac{GJ}{L} & 0 & 0 \\ 0 & 0 & -\frac{6EI_y}{L^2} & 0 & \frac{4EI_y}{L} & 0 & 0 & 0 & \frac{6EI_y}{L^2} & 0 & \frac{2EI_y}{L} & 0 \\ 0 & \frac{6EI_z}{L^2} & 0 & 0 & 0 & \frac{4EI_z}{L} & 0 & -\frac{6EI_z}{L^2} & 0 & 0 & 0 & \frac{2EI_z}{L} \\ -\frac{EA}{L} & 0 & 0 & 0 & 0 & 0 & \frac{EA}{L} & 0 & 0 & 0 & 0 & 0 \\ 0 & -\frac{12EI_z}{L^3} & 0 & 0 & 0 & -\frac{6EI_z}{L^2} & 0 & \frac{12EI_z}{L^3} & 0 & 0 & 0 & -\frac{6EI_z}{L^2} \\ 0 & 0 & -\frac{12EI_y}{L^3} & 0 & \frac{6EI_y}{L^2} & 0 & 0 & 0 & \frac{12EI_y}{L^3} & 0 & \frac{6EI_y}{L^2} & 0 \\ 0 & 0 & 0 & -\frac{GJ}{L} & 0 & 0 & 0 & 0 & 0 & \frac{GJ}{L} & 0 & 0 \\ 0 & 0 & -\frac{6EI_y}{L^2} & 0 & \frac{2EI_y}{L} & 0 & 0 & 0 & \frac{6EI_y}{L^2} & 0 & \frac{4EI_y}{L} & 0 \\ 0 & \frac{6EI_z}{L^2} & 0 & 0 & 0 & \frac{2EI_z}{L} & 0 & -\frac{6EI_z}{L^2} & 0 & 0 & 0 & \frac{4EI_z}{L} \end{bmatrix} \begin{Bmatrix} u_1 \\ v_1 \\ w_1 \\ \theta_{x1} \\ \theta_{y1} \\ \theta_{z1} \\ u_2 \\ v_2 \\ w_2 \\ \theta_{x2} \\ \theta_{y2} \\ \theta_{z2} \end{Bmatrix}$$

[13.10]

Direct stiffness v energy solutions

Strictly speaking, the method just described is called the **Matrix-Displacement Method** (MDM). Alternatively, it might be called the **Direct Stiffness** version of FEA. FEA itself has a great many similarities in terms of calculation steps. However, the MDM is limited to truss and beam elements for which the stiffness matrices may be considered to be exact within the limits of Strength of Materials theory. For more complicated structural components such as plates and shells, direct determination of [k] is not possible. Instead [k] must be derived from energy considerations using for example an assumed displacement function. Because such functions are not exact, a finite element model only approximates to the correct solution. The practical significance of this will be discussed in section 13.3.

Principle of minimum potential and its significance

There are a number of theorems which are useful in developing finite element formulations, one of which is stated below;

Principle of minimum potential:

The total potential (π_p) is given as the sum of the strain energy (U_e) and the potential energy of the load (W). The principle of minimum potential energy states that the change of potential with respect to the displacement is stationary, i.e. $\delta p_p / \delta u = 0$.

The strain energy in an element may be written in matrix notation as:

Background to Finite Element Analysis 267

$$SE = \frac{1}{2}\int_{vol}\{\varepsilon\}^T\{\sigma\}dVol \qquad [13.11]$$

and the potential energy of the external loads applied at the nodes:

$$WD = -\{u\}^T\{p\} \qquad [13.12]$$

In order to make use of this result, it is necessary to assume a displacement function (usually of polynomial form) which must meet a number of conditions,[2]

- The polynomial must contain linear terms in order to be able to represent constant strain and constant terms to admit rigid body motions.
- The number of terms must at least equal the number of degrees of freedom associated with the element.
- The function and its derivatives must be continuous within an element.
- The function *should* aim to provide continuity of displacement and slope between elements.

This continuity is not always possible. C_0 continuity refers to displacements, C_1 to first derivative, etc.

The development of an element stiffness matrix using the 'assumed displacement field' method is shown for the simplest case of a constant strain triangle.[6] Consider a flat triangular plate in plane stress.

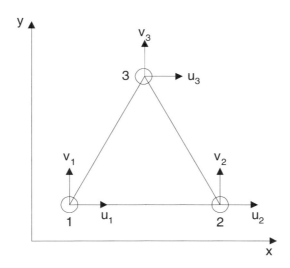

Figure 13.7 *Triangular plane stress element.*

Assumed displacement functions:

$$u = \alpha_1 + \alpha_2 x + \alpha_3 y \quad v = \alpha_4 + \alpha_5 x + \alpha_6 y \qquad [13.13]$$

where α_1 to α_6 denote arbitrary coefficients.

$u_1 = \alpha_1 + \alpha_2 x_1 + \alpha_3 y_1$

$u_2 = \alpha_1 + \alpha_2 x_2 + \alpha_3 y_2$

$u_3 = \alpha_1 + \alpha_2 x_3 + \alpha_3 y_3$ [13.14]

$v_1 = \alpha_4 + \alpha_5 x_1 + \alpha_6 y_1$

$v_2 = \alpha_4 + \alpha_5 x_2 + \alpha_6 y_2$

$v_3 = \alpha_4 + \alpha_5 x_3 + \alpha_6 y_3$

$$\begin{Bmatrix} u_1 \\ u_2 \\ u_3 \\ v_1 \\ v_2 \\ v_3 \end{Bmatrix} = \begin{bmatrix} 1 & x_1 & y_1 & 0 & 0 & 0 \\ 1 & x_2 & y_2 & 0 & 0 & 0 \\ 1 & x_3 & y_3 & 0 & 0 & 0 \\ 0 & 0 & 0 & 1 & x_1 & y_1 \\ 0 & 0 & 0 & 1 & x_2 & y_2 \\ 0 & 0 & 0 & 1 & x_3 & y_3 \end{bmatrix} \begin{Bmatrix} \alpha_1 \\ \alpha_2 \\ \alpha_3 \\ \alpha_4 \\ \alpha_5 \\ \alpha_6 \end{Bmatrix}, \{u_i\} = \begin{bmatrix} A & 0 \\ 0 & A \end{bmatrix} \{\alpha_i\}, \{\alpha_i\} = \begin{bmatrix} A^{-1} & 0 \\ 0 & A^{-1} \end{bmatrix} \{u_i\}$$

[13.15]

$$[A]^{-1} = \frac{1}{2\Delta} \begin{bmatrix} a_1 & a_2 & a_3 \\ b_1 & b_2 & b_3 \\ c_1 & c_2 & c_3 \end{bmatrix}$$ [13.16]

where a_1 to c_3 represent transposed cofactor terms, being differences between products of nodal coordinates x_1 to y_3 and/or 1 and Δ is the area of the triangular element. Equation 13.16 may be substituted into 13.13 to produce $\{u\} = [N]\{u_i\}$ where [N] denotes a matrix of shape functions, $N_i = (a_i + b_i x + c_i y)/2\Delta$ and [N] may be written as:

$$[N] = \begin{bmatrix} N_1 & N_2 & N_3 & 0 & 0 & 0 \\ 0 & 0 & 0 & N_1 & N_2 & N_3 \end{bmatrix}$$ [13.17]

The strains in the element corresponding to equation 13.13 are;

$$\varepsilon_x = \frac{\partial u}{\partial x} = \frac{b_1 u_1 + b_2 u_2 + b_3 u_3}{2\Delta} \qquad \varepsilon_y = \frac{\partial v}{\partial y} = \frac{c_1 v_1 + c_2 v_2 + c_3 v_3}{2\Delta}$$

[13.18]

$$\gamma = \frac{\partial u}{\partial y} + \frac{\partial v}{\partial x} = \frac{c_1 u_1 + c_2 u_2 + c_3 u_3 + b_1 v_1 + b_2 v_2 + b_3 v_3}{2\Delta}$$

or, $\{\varepsilon\} = [B]\{u\}$ [13.19]

$$[B] = \frac{1}{2\Delta} \begin{bmatrix} b_1 & b_2 & b_3 & 0 & 0 & 0 \\ 0 & 0 & 0 & c_1 & c_2 & c_3 \\ c_1 & c_2 & c_3 & b_1 & b_2 & b_3 \end{bmatrix} \qquad [13.20]$$

Clearly the strains are constant over the element since they are independent of x and y, thus the element is called a **constant strain triangle** (CST). Equation 13.19 merely relates the strains to the nodal displacements. Note; the matrix [B] is made up of differences between x coordinates and differences between y coordinates for the three corner nodes.

Having obtained equation 13.20, the next stage is to use the well known stress–strain relationships, such that:

$$\{\sigma\} = [D]\{\varepsilon\} \qquad [13.21]$$

[D] = the material matrix and for an isotropic material:

$$[D] = \begin{bmatrix} \dfrac{E}{1-v^2} & \dfrac{vE}{1-v^2} & 0 \\ \dfrac{vE}{1-v^2} & \dfrac{E}{1-v^2} & 0 \\ 0 & 0 & \dfrac{E}{2(1+v)} \end{bmatrix} \qquad [13.22]$$

The final step is to relate nodal forces corresponding to the above stresses to the nodal displacements and thereby determine the element stiffness matrix.

Equation 13.11 may be rewritten as:

$$SE = \frac{1}{2}\int_{vol} \{u\}^T [B]^T [D][B]\{u\} dVol \qquad [13.23]$$

From the principle of minimum energy,

$$0 = \int [B]^T [D][B]\{u\} dVol - \{p\}$$

and since $\{p\} = [k]\{u\}$

$$[k] = \int_{vol} [B]^T [D][B] dVol \qquad [13.24]$$

This is a more general form than is strictly necessary for finding [k] for a CST, since the thickness is constant and this simplifies the integration.

Higher order elements

The constant strain triangle and the truss element employ linear displacement functions and are known as low-order elements.[2] A higher-order element may be obtained by increasing the degree of the displacement polynomial and placing extra nodes along the element boundaries. When obtaining [k] by direct method, it is necessary to equally space these nodes along straight edges. The three noded constant strain triangle with mid-side nodes would employ displacement functions of the form:

$$u = \alpha_1 + \alpha_2 x + \alpha_3 y + \alpha_4 x^2 + \alpha_5 xy + \alpha_6 y^2$$
$$v = \alpha_7 + \alpha_8 x + \alpha_9 y + \alpha_{10} x^2 + \alpha_{11} xy + \alpha_{12} y^2 \qquad [13.25]$$

and hence the strain varies linearly over the surface. Such an element should give a better representation in areas of high strain gradient and the choice often lies between modelling using many lower-order elements or rather fewer higher-order elements.

Isoparametric elements

This term derives from the idea of mapping (parametric) elements from their real form to an easier to manipulate 'natural' form,[2] such that the mapping functions between global and natural coordinates employ the same shape functions ([N] matrix) as are used to transform displacements from the element nodal displacements.

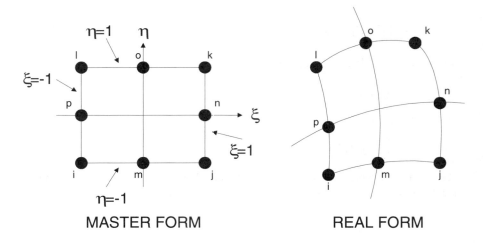

Figure 13.8 *Eight noded quadrilateral element.*

The real value of an isoparametric formulation increases as the complexity of the element, but the mapping concept may be best illustrated by considering a truss element.

If x_i and x_j define the nodal coordinates for the rod, then any point along the length of the rod may be defined using:

$$x = x(\xi) = \frac{1}{2}(1-\xi)x_i + \frac{1}{2}(1+\xi)x_j \qquad [13.26]$$

The element displacement distribution may also be written as a function of the nodal displacements, i.e.

$$u = u(\xi) = \frac{1}{2}(1-\xi)u_i + \frac{1}{2}(1+\xi)u_j \qquad [13.27]$$

or $\{u\} = [N]\{u^e\}$ where $N_i = 0.5(1-\xi)$ and $N_j = 0.5(1+\xi)$ and $\{u^e\}$ is the element nodal displacement vector.

It can be seen that the same shape function matrix [N] is used for interpolation of the element displacements and the geometry. This is a definition of an isoparametric element.

The [B] matrix (see equation 13.19) may be obtained from $[B] = [\delta][N]$. $[\delta]$ stands for d/dx and in order to evaluate [B], the chain rule of differentiation is used since [N] is a function of ξx and not x. Hence $d/dx = d\xi/dx \cdot d/d\xi$. For the bar element in question $dx/d\xi = -x_i/2 + x_j/2 = L/2$. As the Jacobian J is defined as $dx/d\xi$ for a one-dimensional bar, $J = L/2$ and it can be seen that J is a scaling factor since $dx = J\, d\xi$. Further details can be found in reference 2.

Non-linear analysis

A number of yacht structural analyses will need to be non-linear in order to reflect the actual behaviour. Two key non-linearities are those concerning geometry and material characteristics, e.g. large deflection, large strain, surface to surface contact and plasticity/'yielding'.

Consider a strut under load. The compressive axial force will interact with the bending deflections to introduce an additional non-linear bending moment term. Consequently, the [k] matrices discussed to date are to be regarded as initial stiffness matrices.

Non-linearity in the stiffness of structural members may be taken into account by the use of a modifying second stiffness matrix,[7] i.e.

$$[k] = [k_o] + [k_1] \qquad [13.28]$$

$[k_o]$ = a stiffness matrix. This is the usual stiffness matrix if the material properties of the element are constant

$[k_1]$ = the geometrical stiffness matrix which depends on the internal forces in the element and is also known as the initial stress-stiffness matrix.

For the structure we may write:

272   Sailing Yacht Design: Theory

$$\{P\} = ([K_o] + [K_1]) \{U\} \qquad [13.29]$$

The solution of non-linear problems is usually carried out by piece-wise (or step) linear analysis, as follows;

Step 1
Initial stresses are normally zero which means $[K_1]$ = null. Otherwise it is a function of prestresses. An increment of load is applied and the displacements increments are calculated.

Step 2
New material properties are found which correspond to the latest estimate of displacements (i.e. $[K_o]$ is updated). Likewise, geometric/stress effects which correspond to the latest displacements are used to update $[K_1]$.

A second load increment is applied and a new set of increments of displacement is obtained.

The displacements to date are found by adding the new displacement increments to the previous set.

Step 3
Repeat step 2 until all the load has been applied.

As far as the analyst is concerned the main problem in conducting a non-linear analysis involves the selection of the suitable load steps and appropriate convergence criteria.

13.3 FACTORS INFLUENCING THE USE OF FEA

This section is aimed at yacht designers who have or expect to have an occasional requirement to use FEA. It is assumed that the program would need to be run on a PC of modest RAM and hard disk. The comments have less relevance to specialised structural engineering bureaux, shipyard design offices or other experienced users. It is hoped that a discussion of some of the more practical aspects will complement the theoretical explanations described in section 13.2.

The design level: justifying the need for FEA

It is important to realise that results from an FE analysis are only as accurate (at best!) as the accuracy of the load predictions. Load prediction is undoubtedly the weak link in the chain of structural continuity. While it is important not to unduly compound the error by too crude a modelling philosophy, designers need to keep the whole design chain in balance.

The designer must evaluate the cost saving associated with reducing the level of uncertainty which should accrue from using FEA based design, against the potential benefits in terms of improved yacht performance. There are some craft for which a FEA will never be justified, such as non-weight critical boats for which there is a large database of prototypes or poor load definition or where scantlings are based on 'robustness' alone. In estimating the cost of a FEA based design analysis, the designer should appreciate that the work really starts, not ends, when the stress plots are available. These must be interpreted and the structure modified and re-analysed. Time must also be set aside for model verification studies.

Modelling considerations

There are two principal methods of developing finite element models,[8]

- direct modelling
- solid modelling

Direct modelling involves specifying nodal coordinates and element connectivity. This is not as laborious as it sounds since extensive use is made of node and element generation routines. The method is the most time consuming, but provides the user with a good 'handle' on the model.

Solid modelling comes in two forms;

Top down involves the use of 'primitives' such as blocks, spheres etc. which may be 'sculptured' using Boolean operations. Bottom up involves specifying keypoints (like nodes) and generating lines, areas and volumes from these.

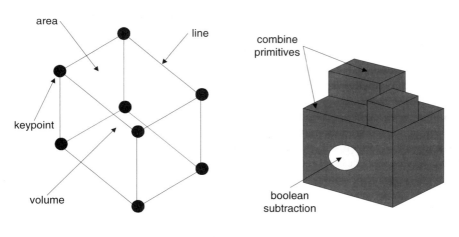

Bottom-up solid modelling **Top-down solid modelling**

Figure 13.9 *Solid modelling options.*

The former method is excellent for creating solid bodies such as an engine block, but many prefer the keypoint based method for thin plated structures. Once the solid model has been developed, the user may specify mesh controls such as element sizes and automatically mesh the body. The process is quite quick, but the designer is a little more remote from the process of discretisation. Solid modelling is the inevitable choice when analysing all but the simplest geometries. The automatic meshing which is associated with solid modelling allows **adaptive meshing**.[8]

Adaptive meshing

Analysts are usually concerned with the question of mesh refinement. It should be remembered that the finite element method is an approximation and it is a problem to know how much refinement should be carried out. It has always been possible to continue to refine the mesh until convergence was obtained using large computers, but this is very time consuming, almost prohibitively so when using direct modelling.

A better way is to allow the computer to calculate energy norms for each element in the model. The energy error is similar to the idea of strain energy. The energy error for a given element (e_i) is obtained from;

$$e_i = \frac{1}{2} \int_{vol} \{\Delta\sigma\}^T [D]^{-1} \{\Delta\sigma\} dVol \qquad [13.30]$$

when $\{\Delta\sigma\}$ is the difference between the average stress vector at node n obtained by averaging the elements connected to node n and the stress vector at node n of the element.

Adaptive meshing is the name given to the process of calculating the error norm and refining the mesh until the error drops below the user defined value. It will be obvious that the process will be invalid at nodes where more than more material type meets, since the error norm is based on average nodal stresses. This is a little unfortunate since yacht designers are often interested in sandwich structures. In addition, care is required for orthotropic materials where properties are referenced to a local (element) x-axis (defined by local I and J nodes); for example, the case of a triangular element generated at a discontinuity.

For the designer using a small PC, hard disk size becomes a problem. For example a eight layer, 900 element composite shell panel model may require many hundred megabytes of storage. An alternative method is to build successively finer meshed, more localised models.

Real and idealised structures

It is perfectly possible to introduce residual stresses, cracks and deviations from plane into finite element models. The spatial variation of material properties may also be included. It is possible to build a very fair reflection of a 'real' boat. However, this requires enormous amounts of data and time. Many node and element generation routines break down when the analyst wishes to introduce random imperfections. Shell warping may also cause some solution related problems. It is easy to see why this process is rarely done and we tend to be analysing 'perfect' structures. This only becomes a problem when designers lose sight of this fact and attempt to 'optimise' an idealised structure as if it were the real thing.

A more serious problem is undue reliance on linear analysis. Stress levels may appear acceptable under a linear analysis, but not so when beam-column effects are included. A finite element analysis is only able to identify failure modes within the limits of the type of solution being performed.

Shell versus three-dimensional 'brick' elements for hull plating

Metal yacht structures will normally be modelled using well behaved quadrilateral isotropic shell elements for plating and stiffener webs and beams/trusses for stiffener flanges. The nodal plane corresponds to the mid-plane of the shell and although it is possible to offset this, the lack of precision in representing plate-stiffener intersection regions is usually acceptable. This is particularly so when the analyst is mainly interested in the distribution of stresses over a substantial portion of the hull.

For sandwich structures, the designer is also interested in the correct representation of through-thickness behaviour. There are four options:

- special layered shell elements
  (six degrees of freedom per node)
- special layered three-dimensional brick elements
  (three degrees of freedom per node)
- a number of brick elements through the thickness (minimum three; face-core-face)
- a number of shell and brick elements through the thickness (minimum three)

The number of nodal planes through the thickness are one, two, four (minimum) and two (minimum) respectively. Given that the designer needs to use a reasonably refined mesh *over* the yacht structure, it can be seen that the choice of option has a significant effect on the model size.

Commercial packages may allow up to 100 layers per element. The analyst may opt to store stresses, strains, failure criteria for each layer or maximum values only. Maximum values will govern pass or fail for a structure, but the designer is starting to lose a feel for the structural behaviour. A balance needs to be struck between being 'swamped' by data and treating finite element as a 'black box'.

Brick elements which are used to represent layers in a sandwich, inevitably have one dimension (through-thickness) which is an order smaller than the other two. Most brick elements perform best when the three sides are similar in size. Interlaminar shear stress performance also varies between the various options.

A final note of caution is required on solid modelling. With a CAD package, if is looks right, it probably is. With FEA this may not be the case. For example, constraints applied to solid models, must be *transferred* to the finite element model (not the same thing at all). A model may look correct, but failure to *glue* two volumes/areas together will mean they are not physically connected in a finite element analysis.

276 Sailing Yacht Design: Theory

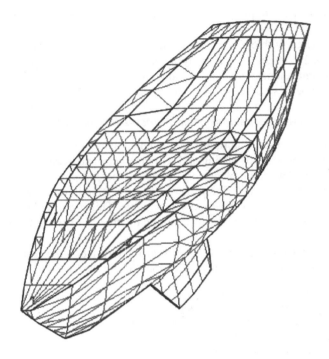

Figure 13.10 *A plausible but incorrect FE yacht model.*

A model may superficially look plausible. However, the stresses predicted around the keel for the model in Figure 13.10, are probably incorrect since the keel has been attached to the shell as if it were 'stuck on'. In fact, it would be necessary to model the keel bolts as pre-tensioned truss elements, surrounding by *contact* elements to reflect the true structural geometry. Contact elements prevent inadmissible displacements such as a keel bolt passing effortlessly through the GRP shell and require a *non-linear* analysis. This of course increases the level of the analysis.

In addition, the model does not contain sufficient elements between major support points to adequately predict tertiary stresses. Normally, one would expect to use five to ten panel elements between frames and stiffeners. Any stresses from such a model would therefore be indicative of primary stresses only. One would also need to question how stringers, for example, have been modelled. It is often possible to adjust material properties and rigidity factors to effectively 'smear' the stiffener over the shell.

13.4 CONCLUSIONS

Finite element analysis is becoming increasingly accessible to the yacht design community due to the availability of easy to use Windows-based applications and more powerful PCs. A commercial FE package may contain over a million lines of coding and represents many man-years of development. It is unlikely that users will ever need to code an FE routine, although familiarity with macros is a considerable advantage when carrying out parametric studies. One might question, therefore, the value of studying the background theory. The answer is clear. All commercial packages contain a disclaimer to the effect that responsibility for the correctness of the results lies entirely with the user. Only by understanding some of the assumptions and limitations involved, can design engineers hope to produce models which not only look right, but give reliable answers. The authors hope this *briefest* of introductions will provide a starting point for engineers who are unfamiliar with the basic concepts of FEA.

REFERENCES

1. *ABS Guide for Building and Classing Offshore Racing Yachts*. 1994.
2. Mottram, J.T. and Shaw, C.T., *Using Finite Elements in Mechanical Design*. McGraw-Hill Book Company, 1996. ISBN 0-07-709093-4.
3. Speer, T.D., *Structural Design of Aluminum Vessels Larger than 50 m*, 2nd Int. Forum on Aluminium Ships, Melbourne, 22-23 Nov .1995.
4. Cook, R.D., *Concepts and Applications of Finite Element Analysis*. John Wiley and Sons, 2nd Edn. 1981.
5. Ross, C.T.F., *Structural Analysis by the Matrix-Displacement Method*. Draughtsmen's and Allied Technicians Association, Richmond, Surrey. 1970–71.
6. Case, J., Chilver, L. and Ross, C.T.F., *Strength of Materials & Structures*.Edward Arnold, 3rd Edn. ISBN 0 340 56829 –1.
7. Ross, C.T.F., *Finite Element Theory in Structural Mechanics*. AUEW publication.
8. ANSYS manual, Vols 1–4.

BIBLIOGRAPHY

Damonte, R, '*Finite Element Analysis of Composites*' Composite Materials in Maritime Structures, Vol 1, Edited by R.A.Shenoi and J.F.Wellicome, Cambridge Ocean Technology series 4, 1993. pp255–279. ISBN 0 521 45153 1

Donaldson, B.K., '*Analysis of Aircraft Structures*' McGraw Hill Int. Ed. 1993. ISBN 0 –07– 112591–4.

CHAPTER 14
MODEL TESTING

A.Claughton
*Wolfson Unit for Marine Technology and Industrial Aerodynamics,
University of Southampton*

14.1 INTRODUCTION

Physical model testing was, until the advent of reliable computational fluid dynamics (CFD) techniques, the fundamental method of research for the development and investigation of sailing yachts. Physical experiments are now used in conjunction with CFD; having anchored the computational results to reliable experimental data, the greater speed and presentational effectiveness of CFD can be used for wider investigations. The task of validating CFD results has led, if anything, to a requirement for greater accuracy and skill from the experimenters, and a diversification into more detailed flow mapping techniques using laser doppler anemometry, hot film probes and flow visualisation. This has arisen from the need to know not just the 'what' about a yachts behaviour but the 'why' of the physical causes. The advent of portable computers has also helped the experimenter in terms of his ability to manage and analyse the experimental process. Analysis of time histories and data, both in the laboratory and at sea have improved the quality of the results. Also advances in machining techniques, transducers and signal conditioning equipment have improved dynamometry and reduced its cost so that one-off experiments are more easily carried out.

To date, in high stakes competitions physical model testing of candidate designs is still regarded as the final arbiter of design choice prior to committing to the build process. This is also the case with racing cars and aeroplanes, despite the undoubted sophistication of their designers and computational techniques. This is not a Luddite claim but an accurate reflection of the current state of the art, where a physical model test is the technique least reliant on assumptions about the physical phenomena at work.

In many situations the designer, and more commonly, the researcher will be faced with a choice between what he would like to do and what he can do, either due to budget and time constraints, or the range of facilities available in his institution. It is important to choose the correct approach to particular aspects of sailing yacht design, and if a good experimental approach cannot be achieved then energy and time might be best expended working with reliable published data. The aim of this chapter is two fold; firstly to identify the parts of the sailing yacht design process that are most amenable to physical testing, and to identify the most appropriate facilities to use, based on experience and reference to typical campaigns, and secondly to describe the test and analysis techniques required to work confidently with these facilities.

14.2 FACILITIES AND APPROACHES

The four primary experimental facilities available in most countries with a maritime infrastructure are:

- 'Large' towing tanks. National facilities capable of accommodating models of more than say 300 kg in displacement (e.g. DERA Gosport Length 258 m, Width 12.2 m, Depth 5.5 m, carriage speed 12 m/s.)
- 'Small' towing tanks. Those tanks, usually associated with academic institutions, capable of towing models of up to 100 kg displacement (e.g. Southampton Institute, Length 60 m, Width 3.7 m, Depth 1.8 m, carriage speed 4.5 m/s.)
- 'Large' Wind Tunnels. National facilities capable of a wind speed greater than 40 m/s with a cross section greater than 4 m^2
- 'Low Speed' Wind tunnels. Facilities adapted for low speed aerodynamic studies such as complete rig testing, flow visualisation etc.

Most aspects of evaluating sailing yacht performance are tractable using the towing tank or wind tunnel. However the use of these 'steady flow' facilities to investigate the unsteady behaviour of yachts usually means that the model tests are often a pre-cursor to full scale testing, either through direct measurements in the natural environment, or by comparative '2 boat testing'. The natural air/water division does not necessarily determine the most appropriate facility for testing. Studies on keel fins and rudders may be carried out using wind tunnels, and towing tanks are occasionally used to look at aerodynamic features of the yacht. However the most common decision faced will be whether to model a complete hull in the towing tank, or some portion of the hull and appendage package in the wind tunnel.

The relative merits of towing tank and wind tunnel tests are listed below.

Table 14.1 Relative merits of towing tanks and wind tunnels

Towing tank	
Pros	Cons
Models the complete picture; hull wave pattern interaction, keel–rudder interaction. Multi element keels can be tested.	Cannot achieve full scale Rn. Forces on keel alone are 10–20% of the total which impairs resolution. Models are time consuming to set up and test. (Typically 1–2 configurations per day).

Wind tunnel	
Pros	Cons
Can achieve 0.5 times or more of full scale Rn.	No modelling of keel induced wave pattern effects.
Rapid testing. (Typically 4–8 configurations per day).	No influence of hull generated waves unless a dummy hull is introduced which may reduce model scale.
Excellent resolution of forces.	
No hull model required for bulb and winglet tests.	Complicated set ups required to model keel fin and rudder interaction, or effect of heel.
Ideal environment for flow visualisation, pressure and velocity measurements.	

If the towing tank is chosen then the next decision is the choice of model size and towing tank facility. This choice is usually conditioned by available budget and time scale. Given unlimited funds and time then a large facility would be preferred to a smaller one, but on some occasions, for example the study of high speed yachts, or for seakeeping studies smaller towing tanks may offer the best technical solution.

The prime function of the towing tank is to predict the full size resistance of a particular hull and appendage combination. In order to determine this the total model resistance must be deconstructed, the viscous component being calculated and subtracted from the total to yield the residuary resistance component. It is the reliable determination of the viscous component that presents the experimenter with the most difficulties, and the motivation for testing large models. The increasing use of large models has arisen from three factors:

1. The uncertainty of modelling hulls with full afterbodies that carry areas of separated flow, such that a small model might reasonably be expected to exhibit a different interaction between wake and wave pattern. This was of particular concern for yachts such as the 12 metres.[1] Milgram[2] cites the state of the art as an ability to use a single form factor (see Chapter 5) derived from upright tests to be used in the extrapolation of heeled data without significant error.

2. A recent trend to appendages with short chords relative to the hull waterline length, and the consequent difficulty of reliably predicting the viscous behaviour of the rudder and keel winglets operating at high lift coefficients.[3]
3. The growth of sailing yacht testing as a viable market for the large towing tanks. Yacht testing until the 1970s was usually the preserve of smaller testing establishments.

Conventional wisdom dictates that, where possible, large-scale testing is carried out, however there are similar considerations of model size in the testing of powered vessels. Ships require large model tests to measure the complex viscous flows around their full sterns and bulbous bows, while faster vessels and planing boats, whose resistance is determined by wavemaking and hull pressure distribution, are routinely tested using 2–3 m models. Consequently the important role that 'small' towing tanks still play in yacht research may be justified by the following considerations;

- Over time the displacement length ratios of sailing yachts has fallen, and thus the waterline length for a given displacement has increased. Thus the 'small' tanks now routinely test models with 2 m waterline lengths, approaching the requirements set out in references 2 and 3.
- Large model testing does not of itself offer greater accuracy, merely the prospect of more reliable viscous behaviour. Dynamometer systems are more difficult to design for satisfactory operation under the loads imposed by a large model, consequently accuracy and resolution are not necessarily improved by using large models, and the converse may be true if an inappropriate towing and dynamometer system are used.
- The debate over the 'Mariner' tests, a 12 m with an immersed transom stern, which did so much harm to the reputation of towing tank testing has been demonstrated to have been due to poor choice of trial horse, rather than an artefact of testing small models, as discussed in reference 1.
- The ease of model construction and the shorter test times mean that tests in 'small' towing tanks can be carried out in about half the time and at half the cost of tests in a 'large' facility. Consequently there should be good technical reasons for shying away from the most economical facility.

Using large models offers the prospect that the behaviour of the boundary layer flows will more closely resemble that on the full size yacht, thereby reducing the scope for error in the scaling of viscous resistance. Also effects of hull shape on viscous resistance may be discerned, and details of keel, hull and bulb interaction may be more reliably determined. Thus the more viscous resistance predominates, or highly loaded short chord appendages are used, then the more the designer is forced towards large models. Conversely for studies where interest is focused on wavemaking resistance, induced drag or centres of lateral resistance then smaller models can be used with confidence.

Reference 4 demonstrates that good correlation can be achieved between 2 and 5 m models of 12 m yachts, and this has been borne out in subsequent tests on IACC yachts. The effects of keel and rudder planform on effective draft (induced drag) can be investigated, as can the behaviour of centreboards and keel winglets. The reliable use of small models is entirely dependent on ensuring consistent and predictable behaviour of the flow over the model. If necessary, flow visualisation techniques can be used to check the behaviour of the aerofoils at model scale.

It should be borne in mind that whatever the model scale, forced transition of the boundary layer will be required to produce consistent results,[5] and this precludes the examination of viscous effects such as the extent of laminar flow on the keel and winglets. Because the towing tank can only yield information about residuary resistance i.e. the effects of volumes and not boundary layer flows, it is often sensible to use sections on the model appendages that differ from those proposed for the full size vessel. For example a model may be fitted with NACA 00 sections while the yacht may sport more aggressive sections with their maximum thickness much further aft (*Sailing Yacht Design: Practice* – Chapter 9). In a similar vein the correct determination of appropriate full scale viscous drag coefficients for the full size appendages is important in predicting accurate full scale drag values. The determination of the effect of detail geometry changes on effective draft, for example through the use of strakes and fairings, is more readily tackled using large models, because these details are more easily modelled, both physically and in terms of their flow conditions. However once attention is turned to details of bulb and wing shape the greater resolution and higher Reynolds Number of the wind tunnel test on fin and bulb alone soon becomes attractive, as described in reference 6. Thus the choice of experimental facility is very much a matter of determining the aspect of the design under investigation and then cutting an experimental coat to suit the budgetary cloth.

For aerodynamic investigations there are no such choices to be made with regard to facilities. There are typically two types of study undertaken, the investigation of mast sections or headfoils and their interaction with the sails, and the study of complete sail plans. The study of masts is most easily undertaken as a two dimensional (2D) test using a standard overhead wind tunnel balance. Many of these are still in existence in commercial and teaching facilities.

This type of testing is relatively easy to set up, and full scale Reynolds Numbers can be achieved. The models are inherently simple to make and a great many configurations can be tested in a single day. These types of test can be augmented with a variety of flow visualisation techniques, ranging from smoke and oil films to more sophisticated particle tracking (PIV) systems or even Laser Doppler Anemometry[7] (LDA). However, employing these complex techniques can increase the test time by a factor of 3 or 4. This type of testing is often supplemented by the determination of pressure mappings[8] and wake circulation.

The testing of complete sailing yacht rigs has always formed an important part of the activities of the yacht researchers at Southampton University. After a period of relative obscurity interest in sail research for racing yachts has experienced a resurgence since 1990. Several wind tunnels now have purpose-built test sections and dynamometry. Used in collaboration with sailors and sail designers these facilities are proving increasingly useful, not only in the development of sail wardrobes for America's Cup and Volvo Ocean Race (née Whitbread Round the

World Race) yachts, but also in the development of multi-masted sailplans for large cruising yachts. The wind tunnel is analogous to the towing tank in that it is able to discern the effects of planform and camber distribution, but because full scale Reynolds Numbers are not achievable some manipulation of the viscous drag component may be required as part of the analysis. Reference 9 describes a series of tests to evaluate sailplans for beach based fishing craft, and is a good example of the broad range of investigations that may be carried out using this type of testing. For testing offwind sails the effect of tunnel blockage on the results must be fully understood, and this often dictates the most appropriate model size.

Sails have perhaps been neglected in terms of experimental study for too long, probably due to the difficulty of marrying the results satisfactorily to performance predictions, and perhaps also due to the lack of handicap rules that allowed some freedom for sailplan development. Published material, e.g. references 10, 11, and 12, is now available and the data analysis and presentation through VPP post processing, such as that described in reference 13, is more accessible and easily assimilated into the practical lexicon of the designers and sailors. This together with updated test techniques using remote control winches and real time displays of forces and moments has opened sailors' minds to 'tuning' the engine as well as trying to reduce the drag of the hull.

Typical campaigns

When evaluating research approaches it is useful to consider the experiences of similar groups. Below are described some typical research programs in which the author has been involved. These have been both high- and low-budget affairs, sometimes involving large multi-disciplinary research teams. In this type of environment the experimental work will occupy a large or small part of the whole effort and will of course be there to lead or support the CFD. Whatever the budget, the program's aims, and partners, it is important to take a reasoned view of what offers the best solution to the technical problems. In some cases technical partners may have different views about what's important. For example car and aircraft designers are focused on down force, or lift slopes, whereas in sailing yachts reducing drag is the key concern. This may mean that software that is excellent in a related field, where the priority areas in terms of performance improvement are different, may be found unsuitable when applied to sailing yacht optimisation.

Whitbread 60

This a closely restricted class (*Sailing Yacht Design: Practice* – Chapter 16) where the main hydrodynamic choice is the waterline beam and stern shape. The towing tank is the obvious facility for developing hull shapes, but there is limited incentive to use large models except for detail appendage work. These yachts are like powerboats and can use similar test facilities. The wind tunnel is useful to optimise the ballast bulb shape which must be correctly shaped to maximise the effective draft, but the towing tank is vital to determine wave-making effects and balance issues. Conceivably the entire program could be done with large tank models, but a more cost-effective solution might be tests on the keel fin and bulb in a wind tunnel supported by 1:8 scale towing tank tests.

The sailplan offers considerable scope for optimisation, not only because there is so much freedom, but also because of the sail limitations that are imposed for the race and each leg. The most successful yachts in the 1997/98 Whitbread race, for example EF Language, all complemented their sail design and development process with wind tunnel tests on complete sail plans.

Open classes

These boats are similar to Whitbread 60's, but have much wider design scope because the rules allow canting keels, twin rudders and complete freedom of rig choice. Towing tank tests at any scale will be useful. The hulls have very low displacement ratios and sail at high Froude Numbers, therefore wavemaking and balance effects dominate the design process.

America's Cup Class yachts

This is a restricted class rule (*Sailing Yacht Design: Practice* – Chapter 16) aimed at producing close racing around a windward–leeward course. The norm is to have well funded campaigns with at least 3 or 4 full time technicians employed; designers are working within a very constrained design space, refinement is the key and small performance improvements can be important, as described in reference 14. Even with only a moderate budget successful R & D programs can be executed using standard experimental techniques and off the shelf computational tools.

In terms of hydrodynamic behaviour these are relatively slow boats and therefore viscous drag is a major consideration, particularly when the complex keel and wing systems are considered. Traditionally large models are used for evaluation and also as the final arbiter in choosing between candidate designs. It is however possible to use 1:8 scale models for canoe body and broad brush appendage studies. Also this small scale of model is suitable for seakeeping tests and to look at unusual ideas without risking budget on potentially fruitless large model tests. The important design decisions are the correct choice of length and displacement, coupled to the optimum waterline beam. Also, the rule offers scope for a variety of bow types, where calm water against seakeeping performance must be judged. The development of keel bulb and wing geometry is ideally suited to testing in a large wind tunnel, where a standard fin can be set up to accept a range of bulbs, which in turn can be used with different wing arrangements. If tandem keels or unusual distributions of lateral area are contemplated then these must be evaluated in the towing tank to confirm wave pattern interference effects.

America's Cup Class Yachts are becoming more stereotyped, keels are at a structural minimum, rudder area is governed by control considerations, and keel wing span is used to 'tune' the boat's performance for specific wind and sea conditions. Because of this a lot of work will be carried out at full scale through careful two boat testing to check how different arrangements behave in real conditions.

The potential aerodynamic investigations are legion. There is scope for optimising the aerodynamic drag of the mast rigging and even the hull itself. In terms of sail aerodynamics there are two distinct considerations, the development of sail force coefficients to ensure that VPP studies closely reflect reality, and the development of new and faster sail designs. Both wind tunnel work and, where time and budget allow, 'sailing dynamometer' tests,[14] can be used for these studies.

IMS and ILC racing yachts

The task of designers here is to beat the IMS formulations, by designing a boat that sails faster than the speeds predicted by the IMS VPP force models. The IMS VPP is tolerably well developed for the type of yachts it deals with but in offering relief to non optimum designs there is scope for exploitation by designers of one-off yachts. One of the target areas is in the rule's calculation of residuary resistance based on the hulls form parameters. This can be investigated with a modest tank program, and higher budget studies may extend to wind tunnel tests on keel bulbs and seakeeping studies, either using experimental or CFD techniques.

14.3 TOWING TANK TEST TECHNIQUES

There are nearly as many techniques to determine the resistance and sideforce of a yacht model as there are towing tanks, witness the numerous contributions in reference 15. The test techniques used to date fall broadly into two categories, 'free' (e.g. reference 16), and 'semi-captive' (e.g. references 4 and 17). In the free system the model is towed from a mast whose height approximates the sailplan centre of effort position, the model is set to a predetermined yaw angle and under the influence of the resulting heeling force the model adopts an equilibrium heel angle that simulates the sailing condition. In order to determine the position of the longitudinal CLR the entire towing mast moves fore and aft under servo motor control until a net zero yaw moment is experienced at the towing position. This approach is elegant and has the merit that every data point relates to a real sailing condition, i.e. the speed, heel and leeway angle will be matched by the full size vessel. There are, however, some drawbacks to this approach, principally that the models must be configured to have the same GM as the real yacht, so ballast keels or a complex moving weight system are required on the model. Also the linkage of the test data points to a specific GM and sailplan centre of effort (CE) height and the cross coupling of heel and leeway make the application of the test results to other sailing conditions difficult. Unlike the 'free' towing system, with a semi-captive system it is entirely possible that none of the tested points, combinations of speed, heel and leeway angle, will match exactly a sailing condition of the full size yacht. Nevertheless, despite this apparent deficiency the system offers methods of analysis that give good insights into the different resistance components and provides a relatively straightforward way to assimilate the data for input to a VPP. Since 1990 most of the commercial testing of sailing yachts has been carried out using the semi-captive technique.

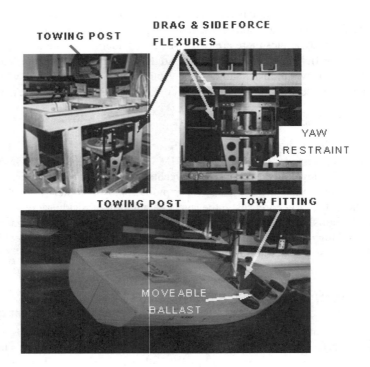

Figure 14.1 *Photograph of a model under test using a single post semi-captive system.*

Figure 14.1 shows a semi-captive test arrangement for 2–3m models developed by the Wolfson Unit. It demonstrates all the typical features of this type of towing mechanism. The model is free to heave, by means of the towing post running on roller bearings, and free to pitch through the gimbal mechanism in the towing fitting. The roll angle is fixed by a restraining arm and pinning quadrant on the aft face of the tow fitting. This restraining arm is strain gauged to measure the restraining roll moment. The towing post is fixed to the yaw adjustment plate by the yaw restraint arm and a yaw moment dynamometer is incorporated into the socket that connects to the tow fitting. The yaw adjustment plate and the model can be rotated bodily under the resistance and sideforce flexures to set the desired yaw angle.

The main pre-occupation in designing a satisfactory yacht dynamometer is the minimisation of interactions between the forces and moments measured by the dynamometer. Each sensor should react to one specific force or moment component and be insensitive to all other components. Most important of these is to ensure that the resistance and sideforce sensors are operating in orthogonal planes aligned with the tank centreline and that there is minimal interaction between the force components. Another potential source of interaction problems is the large hydrostatic roll moments that may be produced by the model as it is heeled over. Often models have a high GM and when restrained at a heel angle, the dynamometer is providing some of the roll moment. This extraneous moment may affect the resistance and sideforce sensors. It is good practice to have sufficient internal ballast in the model so that it can be shifted to roll the model to the desired heel angle, so the roll restraining mechanism and roll moment dynamometers need only resist the hydrodynamically induced roll moment arising from the heeling force. In some systems roll restraint is eliminated altogether and the desired heel angle when under way is set by moving a sliding weight. In the single post system shown in Figure 14.1 the resistance/sideforce interaction and alignment problem is overcome by using large mechanical flexures as the force sensing elements.

One of the drawbacks of the single post system is that it is relatively flexible in yaw, and therefore the set yaw angle will differ from that adopted by the model under the influence of the hydrodynamic yaw moment. Because of the approach adopted to the analysis of the test data this is of little consequence, but of course once larger models are to be towed the single post system becomes less appropriate.

For towing models up to 7 m in a larger towing tank the Wolfson Unit uses a three post system shown in Figure 14.2. In this arrangement the measurement of sideforce is through force blocks on the fore and aft posts, and resistance is measured on the centre post. Through the arrangement of the gimbals on the 3 posts restraint is provided in surge, sway and yaw while isolating the resistance force block from sideforce and vice-versa. Measurement of the hydrodynamic roll moment is through roll restraint beams at the forward and aft ends of the bottom channel.

288 *Sailing Yacht Design: Theory*

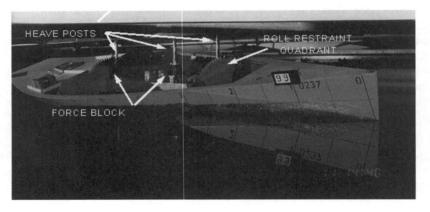

Figure 14.2 *Photograph of a model under test using a 3 post towing arrangement.*

The common themes of these two towing arrangements are:

1. Use of flexures and deflection sensing transducers for force measurement. These are more robust and less sensitive to interaction than strain gauge force blocks.
2. Use of individual force blocks for each measurement. This again helps minimise interactions, and offers the capability to use common components through the system. Of course interaction matrices can be determined for strain gauged force blocks, but the forces on a large yacht model are high and the calibration frame required is complex. The force block dynamometers have in situ calibration mechanisms to facilitate the regular checking.

3. Alignment of the dynamometer on space axes, i.e. with the direction of travel of the yacht.
4. Ability to test on both tacks. Asymmetry can arise both from misalignments in the towing apparatus and misalignments of the appendages. Testing on both tacks allows the extent of these effects to be determined and eliminates them as a concern.
5. Use of a model construction technique and low dynamometer payload systems that allows for at least 50% of the model displacement to be in the form of moveable ballast. This makes the models easy to handle ashore and easy to heel when afloat. Some of the schemes required to allow NC milling often result in models heavier than this, with consequent ramifications in terms of minimising roll moments transferred to the dynamometer.

There are a number of references[4,18,19,20,21] that describe procedures for minimising the systematic errors in the process of tank testing. These fall into two categories, firstly the correct approach to ensure that the model tests mirror the full scale situation, and secondly the method of testing to ensure that random errors are minimised. The errors that can afflict model test results are many and varied. Most of them can be minimised, but not eliminated from consideration, by careful attention to testing protocols and maintaining a high standard of consistency, not only through the physical equipment used but also in the run schedules etc. Published references do little more than hint at the types of problem that may occur. The effect of temperature on viscous resistance is calculable, but may be subject to unforeseen errors or due to temperature stratification in the tank such that appendages and hull experience different flow conditions. From time to time circulation currents may develop in the tank, and these manifest themselves as a virtual modification of the tank axes. This may be linked to temperature gradients in the tank or to circulation shed by the yawed model. Testing the model on both tacks eliminates this to some degree, and is a practical way to be sure that asymmetry effects in the model, or even the dynamometer, are correctly accounted for. Even the time of day, or more likely the effect of continued disturbance of the water can affect the repeatability of tests,[22] hence the need for consistent run schedules model to model. As far as dynamometer and force blocks are concerned, a simple in situ calibration mechanism is a useful feature so that calibration rates can be quickly checked. Systems whose components (Force block, wires, signal conditioning, and A–D) are not routinely dismantled can offer the experimenter calibration rates that do not vary by more than 0.25% over periods of months and even years. Thus even if a rig requires calibration off the carriage the calibration should be done through the same system that is used for the tests.

There are several components to executing a satisfactory model test:

1. Construct an accurate model, with appendage geometries chosen to model volumetric effects while avoiding potential problems with unpredictable viscous flow characteristics.
2. Adopt a proven technique for forcing boundary layer transition on the canoe body and appendages.

3. Devise a consistent procedure to extrapolate the results to full scale.
4. Adopt a test technique that ensures the attitude of the towed model mirrors that of the full size yacht, and also requires minimal restraining moments from the model support apparatus and dynamometer.
5. Determine a matrix of test points that will, once analysed, allow the hydrodynamic characteristics of the model to be determined and the results communicated to performance prediction software.

Scaling

The procedure for extrapolating the results of a model test to full scale are the same as those adopted for conventional ships. The total model resistance is measured, the viscous resistance of the model is calculated and the balance is determined to be residuary resistance. The residuary resistance is extrapolated to full scale using Froude scaling, while the viscous resistance, is extrapolated using some form of model–ship correlation line. The process is shown diagramatically in Figure 14.3, which shows how the viscous component is a greater proportion of the total resistance at model scale than full scale due to the lower R_n. In the early days of yacht testing the characteristic length of the hull was taken as 70% of the waterline length, in order to approximately account for the shorter chord length of the keel, which was an integral part of the hull. Modern yachts have appendages that can be more easily separated from the canoe body, and calculation of viscous drag can be done on the basis of the wetted areas and lengths of hull, keel and rudder. Although there is still some debate about the correct wetted length to use for a hull, provided a consistent value is used at model and full scale this merely manifests itself as a variation in form factor. The model resistance must also be corrected for the tare drag of the turbulence stimulators and the presence of laminar flow on the model. There are a number of arrangements for promoting laminar to turbulent transition, ranging from uniformly spaced studs and sand strips to trip wires. Most laboratories have systems used over a number of years for which they have devised reliable schemes not only for extrapolating the viscous resistance, but also for the calculation of the tare drag of the turbulence stimulators. For fixed geometry turbulation devices a drag coefficient can be used, whereas the use of sand strips requires extrapolation of the data to a zero sand strip width. Further refinements of the extrapolation technique are to use wetted areas and reference lengths that are determined at each speed and heel angle, as described in reference 23. This approach is not necessary for the extrapolation of data on a single hull, but is useful if the analysis is carried out on several hull forms to produce data for the determination of residuary resistance in a general case based on hull parameters.

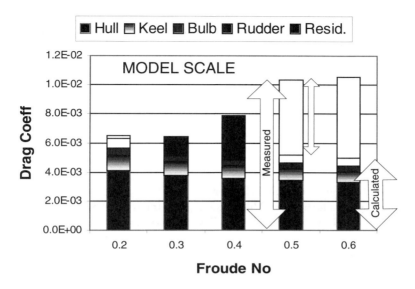
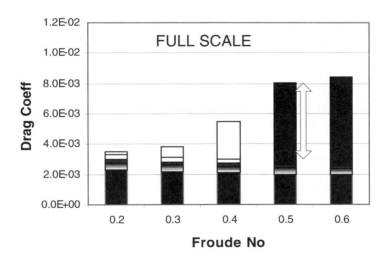

Figure 14.3 *Resistance components at model and full scale.*

The calculation of the viscous component of resistance for the short chord appendages is not peculiar to sailing yachts, as ship experimenters must deal with appendages such as fin stabilisers and shaft brackets, as discussed in reference 24. For modern yacht forms with distinct appendages it is common practice to test both the bare canoe body and the complete model. Comparison of the calculated low speed residuary resistance curves for the two conditions provides a mechanism to check the accuracy of the appendage viscous drag calculation. The only hard and fast rule about turbulence stimulation is that it must be effective, i.e. the transition point must occur in a uniform way run to run, otherwise calculation of a model scale viscous resistance is impossible. The Prohaska plot described in Chapter 5 is the best way of checking that the turbulation method is working, although it gives no information about any modification of viscous effects when the appendages are producing lift.

Test procedure

Test matrix

The polar curve shown in Figure 7.13 contains both the normal polar speed curves and lines of constant heel angle. This figure demonstrates that a yacht can sail at a wide range of speeds and heel angles. As described in Chapter 2 the sailing sideforce depends on the heel angle and CE height, and the leeway is determined by this and the speed and heel angle. With a 'free' model test system the model decides for itself the appropriate combinations, and the experimenter executes a series of tests at different speeds. However, with the restrained model system the experimenter must decide what combination of speeds and heel angles to use to define the behaviour of the yacht. There is inevitably a conflict between selecting test points at the extremities of the operating envelope and copying the actual sailing conditions. This choice will be most often dictated by budget, which will determine how many runs are available, and the use to which the data will be put, For comparison of candidate designs sailing conditions will dominate the matrix, while for analytical studies the conditions may be more widely spread. The type of yacht and the intended race course will also affect the choice of test matrix. For an America's Cup yacht, where racing is confined to windward–leeward courses, the following combination of speeds and heel angles might be chosen :

Table 14.2 Typical test matrix for America's Cup Class yacht

Heel angle	Speed in knots				
15.0		8.5			
20.0	8	8.5	9.0	9.5	
25.0		8.5	9.0	9.5	10.0
30.0			9.0		

In this matrix there is a relatively narrow band of speeds, and higher speeds are associated with higher heel angles, giving a diagonal trend to the matrix.

For a Round the World Race yacht there is a much wider range of speeds and a less strong correlation between speed and heel angle, and therefore a much wider spread of points.

Table 14.3 Typical test matrix for a Round the World Race yacht

Heel angle	Speed in knots		
17.5	8.0	12.0	16.0
25.0	8.0	12.0	16.0
30.0	8.0	12.0	16.0

Testing

Having decided on a test program it is possible to begin the process of testing. The first requirement is the production of an accurate and robust model. Building models with a flat sheer line at a fixed distance above the design waterline is an aid to accurate building and setting up. In order to trim the model when it is afloat it is best to ballast to predetermined weight and use freeboard measurements as a check on trim.

A typical model test will usually begin with a series of tests on the bare canoe body. The hull will be ballasted to the displacement of the canoe body only i.e. minus the appendage volume.

Table 14.4 shows the procedure adopted for each test run, the set parameters are fixed prior to the run and measured during the run.

Table 14.4 Set and measured parameters for a semi-captive yacht model test run.

Set	Measure
Speed	Carriage speed
Heel angle	Drag
Leeway (yaw) angle	Sideforce
Rudder angle	Roll moment
Trim tab angle	Yaw moment
Sail trimming moment	Trim angle
Sideforce tanϕ weight.	Heave displacement.

In order to allow for the absence of the bow down sail trimming moment the model should be trimmed by moving internal ballast. At each speed the moment is $(Drag_{Full\ Scale}/Scale^3)*$ (CE–Towpoint height). The full scale drag can either be determined from a similar vessel or the measured model resistance corrected for the difference between model and full scale viscous drag.

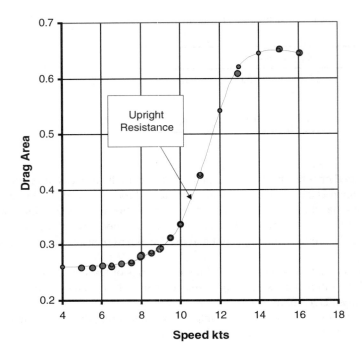

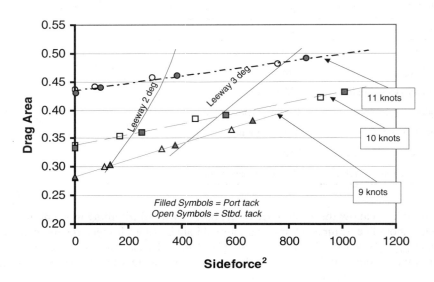

Figure 14.4 *Typical yacht test results, upright resistance (top), heeled and yawed resistance (bottom).*

Once the canoe body only tests are completed the appendages are fitted to the model and an upright resistance test on the complete model can be carried out. A typical resistance curve, expressed as a drag area (A_D = drag/q), is shown on the upper diagram of Figure 14.4. This curve is an expression of the basic resistance quality of the hull and forms the basis of much of the subsequent analysis.

For the heeled and yawed tests, at each speed and heel angle combination a series of tests over a range of yaw angles are carried out. The yaw angles should be chosen so that the sailing sideforce for the heel angle is spanned. Using a predetermined leeway angle sequence that remains the same for all speeds and heel angles will produce a substantial number of test points that are far removed from the 'sailing' condition. Also, for the heeled tests it is convenient to determine the down thrust of the sails (SF tanϕ) based on estimated sailing sideforce for the yacht. This may be applied regardless of the leeway angle set, since it is easier from a practical view point and avoids fluctuations in displacement affecting the determination interpretation of induced drag.

The lower diagram of Figure 14.4 shows typical test data. There is usually some difference in the sideforce for a given leeway on the two tacks and at higher speeds and heel angles a difference of drag tack to tack at the **same sideforce,** *not* the **same leeway,** of 1–2% is acceptable. Greater differences than this indicate a misalignment of the centre planes between the hull and keel and rudder. Typically 4 tests per tack are done at each speed and, although it is tempting from an analytical point of view to make one of these a zero yaw point, there is often some non linearity close to zero sideforce which makes it more useful to begin the tests at some low leeway angle. This process is repeated for all the speed/heel angle combinations in the test matrix.

Analysis

Having determined the data for the test points these are extrapolated to full scale using whatever scaling procedure has been selected. The task now is to use the data to derive some physical insights into the hydrodynamic behaviour of the hull and transmit the results to a VPP. The procedure described here is that adopted by the author, other procedures are described in references 13, 14, 16, and 23.

The first step is to fit a cubic spline curve to the upright resistance results using a least squares fit. This allows the upright resistance to be determined at any speed. Looking now at the heeled and yawed results, usually, unless stall is occurring, a straight line can be fitted to the drag versus SF^2 data points at each speed and heel angle. The slope of the line is determined by the induced drag characteristics of the keel and rudder, combined with the wavemaking effects, although of course bound up in this are the wavemaking characteristics as described in Chapter 5. The slope of the line may be expressed as an effective draft derived from the formula:

$$Te = \sqrt{\frac{1}{\left(\dfrac{dR}{dSF^2}\right)\rho\pi V^2 \cos^2\phi}} \qquad [14.1]$$

$\delta R/\delta SF^2$ = slope of resistance versus sideforce^2 line
ρ = density
V = velocity
ϕ = heel angle

Note that the expression for effective draft differs from that for effective rig height by a factor of $\sqrt{2}$ because effective draft is based on the yacht hull behaving as if it had a reflection plane operating at the water surface.

The intercept of the straight line with the zero sideforce axis (R_0) determines the drag due to heel, and may be expressed as a ratio to the upright resistance R_u. Thus for each tested speed and heel angle the hydrodynamic behaviour can be expressed as an Effective draft (Te) and a heel drag ratio (R_0/R_u), where $R_0 = R_H + R_U$ as defined in equation 7.1, Chapter 7. A plot of typical results are shown in Figure 14.5.

These plots show typical behaviour, in that heel drag increases both with heel angle and speed, and effective draft reduces with increasing speed and heel angle as the keel root comes closer to the water surface.

The other important hydrodynamic features yielded are the yaw and roll moments about the towing axis. The yaw moment results can be analysed to yield the Centre of lateral Resistance (CLR) for each test point. These data must be plotted against sideforce to derive a CLR position at the sailing sideforce. Hydrodynamic roll moment plotted against sideforce yields straight line plots for each speed and heel angle combination. The slope of the line is determined by the distance of the vertical CLR below the roll axis of the fitting, and the intercept with the zero SF axis indicates the stability loss or gain due to the effect of the wave patterns and pressure distribution on the hull surface.

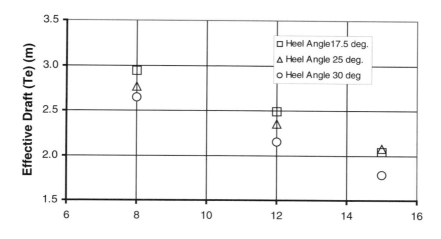
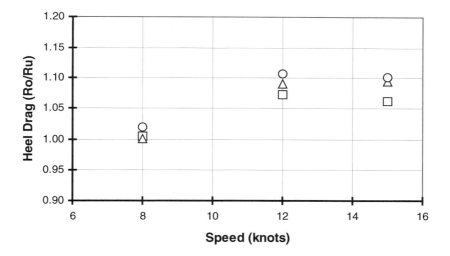

Figure 14.5 *Typical variation of R_o/R_u and Te versus speed for a yacht derived from tank tests.*

Other test techniques

There are a number of related techniques that can help illuminate particular aspects of hydrodynamic behaviour. The use of tank side mounted wave probes to determine directly the energy in the hull wave pattern can give useful insights into the behaviour of different hull types.[25,26]

In the study of appendages different dynamometer systems have been developed where the forces on the keel alone can be measured. This type of study[27] is complex to set up and execute, and of more interest to the scientist than the yacht designer, whose aim is simple, to find out what is 'fast'.

A useful and more easily used technique is the use of pressure tappings in the hull and lifting surfaces, which can be used to look at the incident flow field on keel wings etc. in order to explore the interaction between wing loading and induced drag.

Flow visualisation techniques are an even simpler and more common method of trying to explore the flow mechanisms occurring on the hull and keel. At its simplest paint streak tests can be used to determine local on body flow directions, and to highlight areas of separated flow. Similarly wool tufts mounted on or away from the hull surface can be viewed either through the water surface using a mirror, or via an underwater video camera. This technique is most often used to examine the extent of flow separation over a hull afterbody. When using underwater video cameras, the waves from any surface piercing strut will affect the model, and resistance tests should not be combined with flow visualisation work that involves such struts.

Some towing tanks are equipped with viewing windows in the side and bottom, but it is difficult to get a good view as the model moves past. A useful alternative is the circulating water channel where the model remains stationary and the water is driven past it. This means that tufts and ink bleeds can be viewed at length. Unfortunately these facilities are not ideally suited to resistance testing due to the difficulty of establishing a uniform and level surface on the moving water.

The final form of testing commonly carried out in towing tanks are seakeeping tests. With a few notable exceptions[30] this is confined to tests in head seas, a situation not often encountered by sailing yachts. Nevertheless a conventional wisdom has become established that tests in head seas with the model heeled can be used to assess the added resistance in waves. In power craft seakeeping studies use is made of test in irregular sea spectra, and the statistical properties of the motions are determined. However for sailing yachts, where the focus is on added resistance in waves, then regular wave test are used to derive Response Amplitude Operators (RAO's) as discussed in Chapter 6 and reference 28. This approach relies on the assumption that the motions will remain linear with wave height. Consequently the choice of wave heights for added resistance tests is a fine judgement between keeping the waves small enough to remain in the linear regime while making them large enough to create a reliably measurable amount of added resistance.

14.4 WIND TUNNEL TESTING

Appendages

Two dimensional wind tunnel testing of keel sections has fallen from common use because of the proven reliability of aerofoil section design CFD. Two-dimensional testing of sections may still be required to examine the behaviour of unusual features such as leading edge weed cutter slots. The testing of complete keel, bulb and winglet configurations is still however a common occupation for the sailing yacht designer, because of the complex three-dimensional flows that occur. In this type of study the achievement of full scale Reynolds is not a pre-requisite, as the effective draft is the main deliverable, and an order of merit for Cd_0 can be established using either natural or forced transition at some appropriate point. In conjunction with the force measurements simple oil film flow visualisation techniques can be used to map areas of laminar flow.

The main prerequisite for the satisfactory execution of tests on a yacht keel is the availability of a purpose designed dynamometer. Many wind tunnels still have serviceable overhead balances that were installed for aircraft testing, from which a double model, reflected about the tunnel centreline may be suspended. This is a feasible way to determine the induced drag characteristics of a keel. However the presence of the support struts and the pitch wire require careful attention to be paid to the tare drags. Also this set up gives rise to asymmetric test results due to the affect of the struts on the boundary layer. This can be mitigated to some extent by introducing a dummy pair of struts. The main drawback of this arrangement is that two models of each configuration must be made, and the double model forces a reduction in model scale. However this set up is still preferable to a poorly specified single fin dynamometer.

When testing a single fin cantilevered from the wind tunnel floor or roof the lift forces are typically 200 times higher than the drag forces, and the keel will exert correspondingly high roll moments on the support structure. It is important that the interactions of the dynamometer are well calibrated at this level. To achieve this the wind tunnel dynamometer is often very large, for example, the virtual centre balance in the George Mitchell wind tunnel at the University Southampton, which occupies an enclosure 2.5 m wide and long by 2 m high above the working section. Dynamometers designed for other studies, e.g. measuring forces on buildings, are seldom capable of satisfactory operation under the loads imposed by a keel model. Testing a single model cantilevered from the floor or roof allows the maximum size model to be mounted in the wind tunnel.[6] Details of general considerations relating to this type of testing are provided by Pope and Harper.[29] The major operational difficulty with this type of test is the setting up of an adequate seal at where the keel root meets the wind tunnel roof, or floor. The gap must be small enough to prevent flow leakage, but large enough to avoid fouling of the model under load. This problem is exacerbated when using a keel with a trim tab, because the tunnel roof may be sucked down on to the top of the tab by the low pressure on the keel. As with towing tank tests it is best to use a dynamometer that is aligned with the wind axes. If this is not possible and the whole dynamometer rotates to adjust the incidence angle then great care must be used to ensure that the dynamometer is aligned with the flow at zero incidence, and the model centreline is coincident with the dynamometer centreline. In any event it is important to test both positive and negative incidences to ensure meaningful results, because the body axis forces must be resolved into wind axes to determine lift and drag. Figure 14.6a shows typical

300 *Sailing Yacht Design: Theory*

data for a keel mounted on a body axis dynamometer. It is surprising to see the axial force (Fx) become negative with increasing incidence. Figure 14.6b shows the same raw data resolved into the more useful lift and drag axes. This shows a typical effect whereby the data from positive and negative incidence angles have quite different slopes. This arises from slight misalignments between the dynamometer, tunnel and model centrelines, and can be easily corrected by determining a notional misalignment angle that reduces the separation of the two lines to a minimum, as shown in Figure 14.6c.

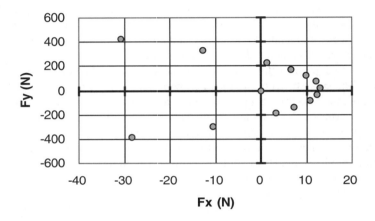

Figure 14.6a *Raw Fx and Fy data from a test on a fin keel.*

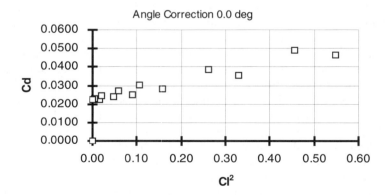

Figure 14.6b *Data resolved to Cd versus Cl^2.*

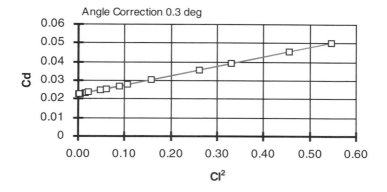

Figure 14.6c *Data corrected for misalignment.*

Testing on an isolated keel fin is quick to carry out and can be extremely useful in deriving orders of merit for different keel arrangements, as shown in Chapter 7. In more sophisticated testing arrangements the presence of the streamline curvature from a dummy canoe body may add a better simulation of the keels operating environment, and at a higher level of sophistication again, a heeled model may be used and the rudder introduced to evaluate the effect of the two lifting surfaces operating in tandem. Compared to towing tank results this type of keel testing is straightforward to analyse. From the Cd versus Cl^2 plot an effective draft and base drag coefficient can be determined. It is relatively simple to transfer this type of result into the VPP as a single value of the basis effective draft, but some other means of determining the speed and heel effects, discussed in section 3, must be found.

Wind tunnel testing of sailing yacht rigs

Facilities

The requirements for satisfactory sail testing do not impose unusual demands on the wind tunnel. There is no requirement for a very high flow speed, nor any need to achieve low flow turbulence levels. The main requirement is that the working section be able to accommodate a suitably sized model. There are however difficulties associated with creating a test section that has wind gradient and twist, both of which are features of the apparent wind experienced by a sailing yacht. In a recirculating wind tunnel it is feasible to grow a vertical velocity gradient over a roughened floor or through blockage elements. If however a vertical variation of wind direction is required then this can only be achieved using an open jet or slatted wall wind tunnel with angled vanes upstream of the model, as discussed in reference 30. While the presence of gradient and twist appears to simulate the sailing condition most closely Figure 14.7 shows that the extent of the velocity gradient and wind twist vary significantly depending on the point of sailing and the yachts speed relative to the true wind, consequently no particular wind field can be regarded as correct.

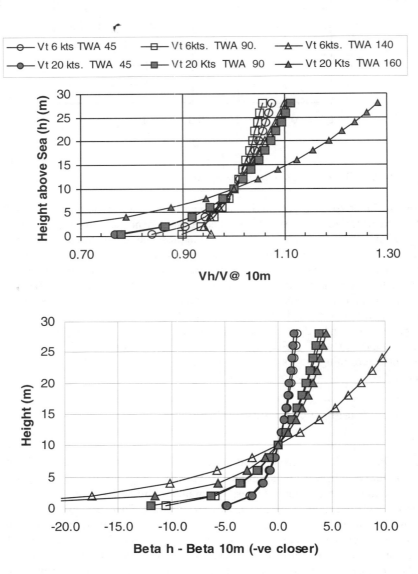

Figure 14.7 *Apparent wind twist and gradient for different true wind speeds (VE) and true wind angle (TWA).*

While devotees of the twisted flow tunnels will not be persuaded to the uniform flow approach it is much easier to determine the correct reference dynamic head in a recirculating tunnel than in an open jet. Also if perfect mimicking of flying shapes is the aim then the influence of cloth weight, particularly for offwind sails, is a consideration, and if the wind speed is set through Froude scaling then the forces are too small to measure reliably. Both open jet and circulating wind tunnels may be used for successful sail development provided the important rather than the esoteric aspects of the test method are correctly addressed.

Test arrangements at the University of Southampton

The first experiments to determine sail force coefficients from wind tunnel tests were carried out by the Southampton University Yacht Research Group.[31,32] Since 1990 initially through work on the safety of sailing vessels[33] the test equipment and methods have been completely revised as described in references 10, 11 and 12. The test arrangement used in the 4.6 m (wide) by 3.7 m (high) low-speed wind tunnel at Southampton University is shown in Figure 14.8.

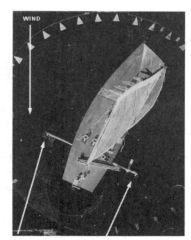

Figure 14.8 *Sailing yacht model in the University of Southampton Wind Tunnel.*

A complete model of the hull, mast and rigging is mounted on a six component dynamometer which is fitted under the wind tunnel floor. The dynamometer rotates with the model so that apparent wind angles between 0–180° can be simulated. The forces are measured track axes, that is the driving and heeling forces are measured and displayed directly during the tests. These forces are transformed to the wind axis system for the derivation of lift and drag coefficients. The model centreline may be set at an angle to the Fx axis to simulate the effect of leeway directly. The turntable rotates with the model, but is independent of it with the model support links from the force blocks passing through slots in it. The underwater part of the hull fits into a water-filled trough in the turntable which provides a frictionless seal and simulates the water surface. In this way no tare forces from the turntable are measured.

Immediately upstream of the model are two fine mesh screens, the primary purpose of which is to produce a uniform air flow in the test section, but the screens also have the beneficial effect of creating fine scale turbulence in the airflow. This increases the effective Reynolds Number and hence improves the modelling of leading and trailing edge separation zones. It is important that these separation zones be representative of full scale conditions since the sails, particularly offwind sails operate at high lift and drag conditions with substantial areas of flow separation.

The wind speeds used for the tests are usually in the range 5–7 m/s which produces forces that match with the dynamometer capabilities. The higher speed is used for upwind sails and reduced speeds for tests on offwind sails to keep the sheet and rigging loads from overloading the sheet winches or causing structural damage to the model.

Models

Model size, particularly for offwind sails is dictated by the need to keep wake blockage corrections at a reasonable level, and to a lesser degree by the availability of off-the-shelf spars and fittings. Using a sail area that is 10% of the tunnel cross section the correction is approximately 25% of the measured forces and can be reliably applied.[29,34,35] The model of the hull should be representative of the yacht, but need not be absolutely accurate since the hull windage can be measured independently from the sail forces. Model yacht sail winches are fitted into the model and these are operated remotely by the sail trimmer/experimenter in the control room who can see the model through large glass windows. The mast section should approximate the full-size dimensions and the rigging arrangement should model the real yacht, as many sail sheeting arrangements are dictated by rigging and spreader positions. The tuning of the mast to match sail shape is identical to that adopted full size. The model sails are cut from mylar film or other light laminated sail cloth, using scaled versions of the mould shapes. Fitting out of the hull models falls between the requirements of dinghies and model yachts. If model yacht techniques are used then the model is too flimsy to withstand the loads imposed from the sails, but if sailing dinghy lines and fittings are used then these are often so large that correctly scaled clearances between the sails and the deck cannot be achieved.

Test procedure

Typically tests in a wind tunnel are carried out without any reference to the performance of the yacht, while when sailing, sail forces cannot be measured directly but can be inferred from changes in boat speed. The wind tunnel operates at constant velocity, removing the fluctuations in wind speed and direction that make full scale trials so difficult, but the trimming of the sails to control heeling force as wind speed rises must be simulated by the experimenter since the rig is rigidly attached to the dynamometer at a predetermined heel angle.

Models may be tested at any heel angle, but practical considerations dictate that only one or two are used. VPP programs typically use upright sail force coefficients and make the correction for heel through the apparent wind velocity triangle, so if tests are carried out on a heeled model it is necessary to correct back to the upright condition. For upwind sails the gap between the sails and the deck does not change significantly with heel angle, because the sails are sheeted close to the hull centreline. However, for reaching sails the foot of the sail operates close to the sea surface and is affected by heel angle.

The sail forces and moments from the dynamometer are displayed in the control room. The sail trimmer uses a 'real time' display to assist in achieving optimum sail settings, but once a satisfactory trim has been achieved then the sail controls are left fixed and force data is acquired over a 20 second period to determine a single data point.

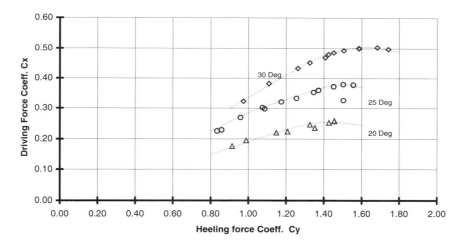

Figure 14.9 *Raw wind tunnel data Fx versus Fy at three apparent wind angles.*[11]

Figure 14.9 shows typical wind tunnel results for a sloop rig with mast head genoa at three different apparent wind angles. The data are plotted as driving force against heeling force coefficient. At each apparent wind angle the data were obtained by first optimising the sail settings to achieve the highest possible driving force. The data also show that by over sheeting the sails driving force may reduce at the highest heeling forces. Having maximised the driving force the sails are then adjusted to reduce the heeling force, with a minimum reduction in driving force. The reduction in heeling force is achieved by first by easing the main sheet, thus allowing the mainsail to 'twist' more and reducing the amount of separated flow. To reduce heeling force further the main sail traveller and genoa sheet must be adjusted to reduce the angle of attack of the sails. This reduction in heeling force is necessary in stronger wind conditions to keep the yacht sailing at its optimum heel angle.

Figure 14.9 shows that for the same heeling force the driving force is higher at greater apparent wind angles. The envelope curves are drawn through the test points with the highest driving force for a given heeling force. During the testing process it is possible to produce 'non optimum' settings which fall below the envelope curve. These are a reflection of real life behaviour where one sail trimmer may do better than another.

The lines drawn from the origin tangentially to the envelope curves define the sail setting with the maximum driving force:heeling force ratio. This is the point where the sails are operating most efficiently, but in real sailing conditions the heeling force coefficient at which the sails are set is governed by the apparent wind speed and heel angle of the yacht. In light winds the sails will be trimmed to produce maximum driving force, as the wind rises the sail will be eased to control heel angle. As the wind rises the sails will be eased through the 'optimum' point down to the lowest heeling force coefficient, where again the driving force heeling force ratio is once again less than optimum. The sail trimmers skill is to maximise the driving force at the heeling force dictated by the yachts heel angle.

Data analysis

The presentation of driving force and heeling force is most easily related to a sailor's frame of reference, but more analytical information can be extracted from the data by transforming it to lift and drag coefficients (Cl and Cd). Since both the induced drag and the profile drag vary with the square of the lift it is informative to plot Cd versus Cl^2.

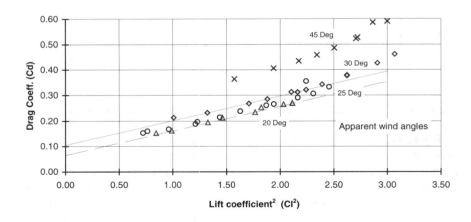

Figure 14.10 *Cd versus Cl^2 plots from wind tunnel tests on a typical sloop rig.*

Figure 14.10 shows the transformed data from Figure 14.9. Lines have been drawn through the data for each apparent wind angle to pass through the lowest drag values for each value of lift. Data that lie above the lines are non optimum sail settings, except at higher values of lift coefficient where, due to the growth of separated flow over the sails as they are sheeted for maximum drive, the drag coefficients rise above the straight line. The behaviour of a particular sailplan can therefore be characterised by four features derived from the Cd versus Cl^2 plot:

1. The slope of the straight line, which is commonly expressed as an effective rig height (He)
   $He = \sqrt{(A/\pi(\delta Cd/\delta Cl^2))}$
2. Cd_0 – The intercept of the straight line with the zero lift axis. The variation of Cd_0 with apparent wind angle can be related to the windage drag of the bare mast and rigging determined from tests without the sails.
3. The maximum achievable Cl at each apparent wind angle.
4. The Cl value above which separated flow occurs.

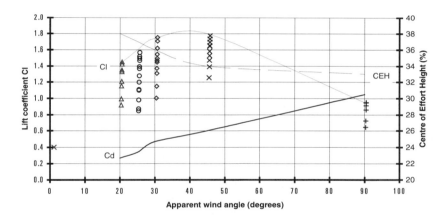

Figure 14.11 *Envelope curves from tests on a single sail combination.*

Figure 14.11 shows how it is possible to characterise a particular combination of sails by the envelope curves, of maximum Cl and corresponding Cd, the effective rig height and Cd_0 plotted against apparent wind angle. The wind tunnel results also yield the vertical and longitudinal position of the centre of effort (CE). For complex sail plans it is necessary to derive envelope plots like Figure 14.11 for different combinations of sails not only over different portions of the apparent wind angle range, but also for sails that are configured for different wind strengths, e.g. reefed sails. Thus, while the trimming of sails for optimum conditions is quite time consuming, ultimately simple parameters can be derived to compare the performance of different sails and sail plans.

REFERENCES

1. Ward, P. and Savitsky, D. *Some correlations of 12 Meter model test results*, Proceedings of ITTC, 1987.
2. Milgram, J.K., *Fluid Mechanics for Sailing Vessel Design*, Annual Review of Fluid Mechanics, 1998.
3. Kirkman, K.L. and Pendrick, D.R., *Scale effects in sailing yacht hydrodynamic testing.* Transactions SNAME, 1974.
4. Campbell, I. and Claughton, A., *'The interpretation' of results from tank tests on 12 m Yachts*. Proceedings of the eigth Chesapeak Sailing Yacht Symposium (CSYS), SNAME. 1987.
5. Larsson, L. *Scientific methods in yacht design.* Annual Review of Fluid Mechanics, 1990.
6. Tinoco, E.N. *et al, IACC Appendage studies.* Proceedings of 11th CSYS, SNAME, 1993.
7. *Applications and accuracy of LDV measurements*, Transactions of 18th ITTC, Volume 2, 1987.
8. Wilkinson, S., *Static pressure distribution over 2D mast/sail geometries.*, Marine Technology, Vol 26, No. 4, Oct. 1989.
9. Marchaj, C.A., *The comparison of potential driving force of various rig types used for fishing vessels*, Proceedings of 8th CSYS, SNAME, 1987.
10. Claughton, A.R. and Campbell, I.M.C., *Wind Tunnel Testing of Sailing Rigs,* 13th International Symposium on Yacht Design and Construction (HISWA), 1994.
11. Campbell, I.M.C., *Optimisation of a sailing rig using wind tunnel data.*, Proceedings of 13th CSYS, 1997.
12. Campbell, I.M.C., *The performance of offwind sails obtained from wind tunnel tests.* International Conference on The Modern Yacht. RINA, 1998.
13. Claughton, A.R. and Oliver J.C. *Development of a multi-functional velocity prediction program (VPP) for sailing yachts.* International Conference CADAP95, RINA 1995.
14. Milgram, J.K., *Naval architecture technology used in winning the 1992 America's Cup match.*, Transactions Society of Naval Architects and Marine Engineers, (SNAME), 1993.
15. Proceedings International Towing Tank Conference (ITTC), Advances in Yacht Testing Techniques, 1987.
16. van Oossanen, P., *Predicting the speed of sailing yachts.* Transactions SNAME, 1993.
17. Gerritsma, J., Keuning, J.A. and Onnink R, *'Geometry resistance and stability of the Delft Systematics Yacht Hull Series'.* International Shipbuilding Progress, Vol 28, No. 328, Dec. 1981.
18. DeBord, F. and Teeters, J.R., *'Accuracy, Test planning and Quality control in sailing yacht performance model testing'*, Proceedings of New England Sailing Yacht Symposium,, SNAME, 1990.
19. Kirkman, K.L., Written contribution, 20th ITTC Yacht testing session, 1993.
20. McRae, B., Binns, J., Klaka, K. and Dovell, A., *Windward performanceof the AME CRC Systematic Yacht Series.* Preceedings International Conference on The Modern Yacht, RINA, 1998.
21. Parsons, B.L. and Pallard, R., The Institute for Marine Dynamics, *Model Yacht dynanometer*, 13th CSYS, SNAME, 1997.
22. Talotte, C., Delhommeau, G., Kobus, J.M., *Adaptation of experimental and numerical results.* Draft manuscript, Ecole Centrale de Nantes, 1995.

23. Teeters, J.R., *Refinements in the technique of tank testing sailing yachts and the processing of test data*. Proceedings of the 11th CSYS, SNAME, 1993.
24. Kirkman, K.L. and Kloetzli, J.W., *Scaling problems of model appendages*. Proceedings of the American Towing Tank Conference, 1980.
25. Insel, M., Molland, A.F., *An investigation into the resistance components of high speed displacement catamarans,* Transactions, RINA, Vol. 134, 1992.
26. Binns, J.R, Klaka, K. and Dovell, A., *Hull-Appendage Interaction of a sailing yacht, investigated with wave cut techniques*. Transactions of 13th CSYS, 1997.
27. Keuning, J.A., Binkhorst, B-J., *Appendage resistance of a sailing yacht hull.*, Transaction of 13th CSYS, 1997.
28. Thomas, G., Klaka, K and Dovell, A., *An investigation of windward performance of yachts in waves*, International Conference CADAP95, RINA, 1995.
29. Pope, A. and Harper J.J., *Low speed wind tunnel testing.*, Wiley ISBN 0471 69392 8.
30. Flay, R.J., Jackson, P.S. *Flow simulation for wind tunnel studies of sail aerodynamics*. 8th International Conference on Wind Engineering, Londan, Ontario, 1991.
31. Marchaj, C.A., *A critical review of methods of establishing sail coefficients and their practical implications in sailing and in performance prediction*, Wolfson Marine Craft Symposium, University of Southampton, 1978.
32. Marchaj, C.A. and Tanner, T. *Wind tunnel testa on a 1/4 scale Dragon ring*, University of Southampton Yacht Research, Report No. 14.
33. Deakin, B., *Model test techniques developed to investigate the wind heeling characteristics of sailing vessels and their response to gusts*. Proceedings of 10th CSYS, SNAME, 1991.
34. Ranzenbach, R. and Mairs, C., *Experimental determination of Sail performance and blockage corrections*. Proceedings of 13th CSYS, 1997.
35. Marchaj, C.A., *Wind tunnel blockage effects on a 'soft' sail at incidence angles beyond stalled flow (up to 170°) and corrected data concerning the Finn type rig*. University of Southampton Yacht Research Report No.35.

CHAPTER 15
SAFETY ENGINEERING

B. Hayman
Det Norske Veritas AS

15.1 INTRODUCTION

The concept of 'safety engineering' is relatively new to sailing yacht design. In this chapter the theoretical considerations of safety engineering are described, from deterministic approaches to probabilistic methods.

The discussion of safety engineering will first of all focus on the structural design process. Ways of dealing with other important safety aspects, such as collision, fire and explosion risks, will also be briefly addressed.

15.2 BASIC CONCEPTS

A basic acceptance criterion for structural design

Traditional structural design has generally been based on establishing a single, representative value of each load applied to the structure, analysing the structure to find details of the induced stresses and deformations, and comparing these with acceptable values. For stresses, these acceptable values are often a measure of the strength of the material (e.g. an estimated lower bound yield or ultimate tensile strength for the case of tensile stress). For loadings that may cause buckling, the allowable stress may be a function of the geometrical properties of the structural element concerned, as well as the basic material properties. For cases in which two or more separate loads may occur simultaneously, some way of combining either the loads or their effects has also to be included in the process.

In its simplest form, the value of load assumed may be an upper bound value for the load in question minus the greatest value of load that could possibly be imagined for the configuration considered. Similarly, the allowable stress may be a lower bound value for the strength of the material. Assuming that the maximum induced stress is a simple, monotonically increasing function of the load, the acceptance criterion is then

$$S_{max} < R_{min} \qquad [15.1]$$

S_{max} = stress due to upper bound load
R_{min} = lower bound strength (allowable stress)

Provided the structure is so simple that (a) it is certain to be constructed in exactly the way assumed in the design, and (b) the structural analysis can be performed with high precision and no uncertainty, this inequality ensures safety without the need to introduce any further safety factors into the design process. This situation is illustrated schematically in Figure 15.1(a).

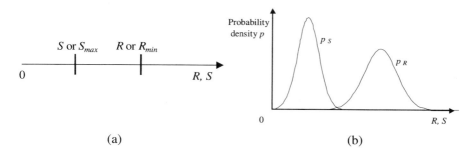

Figure 15.1 *Illustration of relationship between load-induced stress S and strength R (a) when the values of both S and R (or the maximum S and minimum R) are known with certainty, and (b) when they are uncertain.*

In reality, the situation is generally more complicated than this because

- the applied loads often have a so high uncertainty that it is impossible to set a maximum value, or, even if one could do so, a design based on such a maximum value would be impracticable or prohibitively expensive; this is especially the case with environmental loads;
- the values of stresses that are calculated from the applied loads are subject to uncertainties due to possible deviation of the as-built structure from the one assumed in the design and approximations, and to other simplifications and inaccuracies in the structural analysis;
- there is always a possibility that the material or structural element strength will be below the assumed minimum value.

The presence of these uncertainties can be illustrated by Figure 15.1(b), in which the probability density for the load-induced stresses and that for the strength are compared. The overlapping 'tails' of these distributions illustrate the finite probability that S will be greater than R and that the structure will be unsafe, however much we try to ensure that $S < R$.

In structural design it has been necessary to introduce safety margins in the design calculation process to account for uncertainties. In traditional design this has often been done simply by specifying an allowable stress that is appreciably lower than the estimated lower bound material strength. In such cases the allowable stress has generally been set at a value that is considered, on the basis of experience or judgement, to give an acceptably low likelihood of failure. As the allowable stress is the nominal strength reduced by a factor to allow for uncertainties, this approach is referred to as **stress factor design**. An alternative approach that has been used, especially in cases when the failure modes considered involve global collapse of the structure (such as in the plastic design of rigidly jointed steel frameworks), is to require that the applied load, when augmented by a given safety factor, should not exceed the estimated collapse load. This is referred to as **load factor design**. In fact there are several other ways in which uncertainties can be accounted for in the design of structures so as to give an acceptable level of safety.

Uncertainties

There are many types of uncertainty that need to be accounted for in structural design. These concern

- Individual applied loads (sometimes more generally referred to as 'actions')
- The way in which individual loads need to be considered in combination with each other, e.g. to what extent the least favourable load of one type occurs simultaneously with that of another type
- Deviation of the actual as-built structure from that assumed in the design and analysis
- Accuracy of structural analysis used to find stresses and deformations from given loads
- Effects of these stresses and deformations in terms of failure of the structure or structural element (particularly when the potential failure involves buckling)
- Material properties (both strength and stiffness)

Uncertainties are sometimes divided into four broad categories, though some uncertainties do not fall clearly into one single category:

- **Physical uncertainty** (also known as **intrinsic** or **inherent** uncertainty) is a natural randomness of a quantity, such as the yield stress of a material as caused by production variability, or the variability in wind and wave loading.
- **Measurement uncertainty** is uncertainty caused by imperfect instruments and sample disturbance when observing a quantity by means of some form of instrumentation system.
- **Statistical uncertainty** is uncertainty due to limited information such as limited number of observations of a quantity.
- **Model uncertainty** is uncertainty due to imperfections and idealisations made in physical model formulations for load and resistance, and choices of probability distribution types to represent uncertainties.

Two other terms are sometimes used in making what is often an important distinction between types of uncertainty:

- **Aleatory** uncertainty, which represents natural randomness of a variable and corresponds to the 'physical uncertainty' listed above. Such uncertainties cannot be reduced in the design/analysis process. (Uncertainty in material strength might be reduced by changing to an alternative material supplier or changing the supplier's production process, but in the present context this is considered outside the control of the designer/analyst.)
- **Epistemic** uncertainty, which represents uncertainty in a variable due to lack of knowledge. This includes measurement, statistical and model uncertainties as defined above. Such uncertainties can be reduced by, for example, increasing the number of observations or improving the measurement or calculation method.

The main uncertainties in the design of sailing yachts will concern:

- *Physical uncertainty in the loads*, e.g. local slamming loads on the hull structure
- *Physical and statistical uncertainties in material properties*, particularly for fibre reinforced composites (see also Section 15.6)
- *Modelling uncertainties*, particularly if simplified methods of analysis are used in the structural design

Ways of dealing with uncertainty

There are several ways in which uncertainty can be taken care of in the design process. These can be divided into two very broad categories:

- Methods based on experience and judgement. Many older structural design codes and standards, and also parts of ship classification rules, are of this type. There are two major sub-categories:
    - rules in which dimensions are given directly in terms of formulae that have been found to give acceptable results in practice;
    - design codes for structures in which stresses and deformations are assessed but in which the allowable stresses are values that have been found to give acceptable results in practice.

  In both the above cases, events involving failure of a structure may from time to time lead to a re-evaluation, and adjustment, of safety factors.

- Methods based on probabilistic approaches. These try to ensure that the probability of structural failure is less than a specified value. Such methods are generally described as 'reliability-based'. Reliability is defined as the probability that a structure will not fail during its specified life-time, i.e.

$$P_R = 1 - P_F \qquad [15.2]$$

P_R = the reliability (i.e. the survival probability)
P_F = the failure probability

Limit states

The concept of **limit states** is an essential element in modern structural design methods and in the formulation of probabilistic approaches. A limit state may be defined as a state beyond which a structure no longer satisfies the requirements put to it. Commonly in design codes and guidelines, such as ISO 2394[1] and Eurocode 1,[2] limit states are divided into categories such as:

- **Ultimate limit states**, which generally correspond to the maximum load carrying capacity of the structure and are therefore directly related to safety of the structure. Such limit states include
  - loss of static equilibrium of the structure considered as a rigid body (i.e. overturning or capsizing),
  - rupture of critical parts of the structure because either the ultimate strength (in some cases reduced by repeated loading) or the ultimate deformation of the material is exceeded,
  - transformation of the structure into a mechanism.

- **Serviceability limit states** which correspond to criteria related to the ability of the structure to perform its function as intended. Such limit states may be related to
  - deformations which affect the use or appearance of the structure,
  - excessive vibrations causing discomfort or alarm to users of the structure or affecting non-structural elements or equipment mounted on the structure,
  - local damage such as cracking which may reduce the durability of a structure.

The two categories described above are commonly used for all types of structures, including buildings, bridges, offshore structures and ships. Their principal distinction is related to the *consequences* of the structure's exceeding the limit state. In the one case the structure's safety is in jeopardy and the structure itself (or parts of it) may be lost with consequent loss of life and property. In the other the structure's functionality may be lost, but there is no immediate danger of loss of life or property; normally it will be possible to take some corrective action so that the structure's functionality can be restored.

In some cases it is convenient to make other distinctions, setting up separate categories, for example, for certain types of failures and their consequences. Thus DNV's class note on reliability analysis of marine structures[3] includes two separate categories:

- **Fatigue limit states**. These have a special nature because the initial onset of fatigue damage has much in common with a serviceability limit state – it may affect the structure's functionality but will generally not lead to immediate collapse. Provided the fatigue damage is detected soon enough, repairs can be carried out that will restore the structure to its original condition. However, if fatigue damage is allowed to progress unhindered it may eventually lead to degeneration of the structure so that it reaches an ultimate limit state.
- **Progressive collapse limit states**. These are a special case of ultimate limit states, in which the structure is assessed in a damaged condition. This is to

ensure that a limited local failure or damage will not lead to immediate failure of the entire structure.

The concept of limit states is useful because it encourages the designer to consider systematically the conditions under which the structure might fail and the modes of failure involved, and provides a framework for doing this. The limit state categories are important because it is natural to require a lower failure probability for a limit state for which the consequences are severe than for one having relatively mild consequences. In terms of simple safety factors this is the same as saying that the safety factor against a type of failure having severe consequences should be greater than that for a type of failure having milder consequences.

Basic variables in structural design

Basic variables to be considered in a design are of three types:

- **Loads**. A **load** is defined as any action that causes stress or strain in the structure. In some codes the word **action** is preferred to **load**. Symbol F.
- **Material properties** such as ultimate tensile, compressive or shear strength, yield strength or modulus. Symbol f.
- **Geometrical parameters** that describe the shape, size and arrangement of a structure or structural element. Symbol a.

From these, two other types of parameter can be derived:

- **Load effect**: The effect of a single load or combination of loads on the structure, such as stress, stress resultant (shear force, axial force, bending moment, torque), deformation, displacement, motion, etc. Sometimes referred to as **action effects**. Symbol S.
- **Resistance**: The capacity of a structure or part of a structure to resist load effects. Symbol R.

15.3 FOUR LEVELS OF STRUCTURAL RELIABILITY METHODS

Introductory remarks

The brief introduction to structural reliability presented here is partly based on the comprehensive text book by Madsen, Krenk and Lind.[4] This, or the later book by Ditlevsen and Madsen[5] should be referred to for further details. A more elementary introduction to the subject is provided by Thoft–Christensen and Baker.[6] Some ideas have also been taken from the more recent work by Skjong et al.[7]; this explores in particular the relationship between structural reliability analysis and quantitative risk analysis, which is briefly addressed in Section 15.8.

Level I – Deterministic reliability methods

In a level I reliability design, the design calculations are performed deterministically rather than probabilistically and each variable is represented by a single value. The acceptance criterion may be expressed in a simple, allowable stress or load factor format. Alternatively, a limit state format using partial (multiple) safety factors may be used, as in many modern structural design codes for land-based and offshore structures. An advantage with the partial safety factor format is that safety factors can be constructed in a systematic way that accounts for each source of uncertainty. The partial safety factor approach is described in the following.

Each of the basic variables, i.e. loads (actions), material properties and geometrical parameters, is in general a stochastic variable. The first step in the partial safety factor approach is to represent each stochastic variable by a single, nominal value referred to as a **characteristic value**. The characteristic value is normally defined as a fractile of the probability distribution of the variable. The term **representative value** has sometimes been used instead of characteristic value. Here the characteristic values of load, material properties, geometrical parameters, load effects and resistance will be represented by symbols F_k, f_k, a_k, S_k and R_k respectively.

Often the characteristic values of strength are defined as those corresponding to either the 5% or the 2% fractile (i.e. having a 95% or a 98% probability of being exceeded) which, for a normally distributed variable, is the mean minus either 1.64 or 2.05 standard deviations, respectively. For fatigue strength it is common to use either the 2.5% fractile or the mean minus two standard deviations, these being very nearly the same. For material strength lower bound values are also sometimes used. For loads (or actions) the characteristic value is commonly defined as the 95% or 98% fractile, i.e. the value having either a 5% or a 2% probability of being exceeded (equal to mean plus either 1.64 or 2.05 standard deviations for a normally distributed variable) though for environmental loads on ships etc. other statistical measures are also used. Table 15.1 shows the number β of standard deviations above or below the expected (mean) value that corresponds to a given fractile.

From the characteristic values of the basic variables we define a set of *design values* of these same variables:

$$F_d = \gamma_f F_k \qquad [15.3]$$

$$f_d = \frac{f_k}{\gamma_m} \qquad [15.4]$$

$$a_d = a_k \pm \Delta a \qquad [15.5]$$

γ_f, γ_m, and Δa are factors to allow for uncertainty in F, f and a.

From these values, design load effects S_d and design resistances R_d are calculated. In this calculation additional modelling uncertainties have to be accounted for. This is done by means of factors γ_{Sd} and γ_{Rd}. The acceptance criterion can then be written

$$\gamma_n S_d(F_d, a_d, \gamma_{Sd}) \leq R_d(f_d, a_d, \gamma_{Rd}) \qquad [15.6]$$

γ_n = a coefficient to adjust the target reliability level to account for the importance of the structure and the consequence of failure.

Table 15.1 Number of standard deviations from mean value for a given fractile (normal distribution)

% fractile at β standard deviations		No. of standard deviations (β)	% fractile at β standard deviations		No. of standard deviations (β)
Below mean	Above mean		Below mean	Above mean	
15.9	84.1	1.00	0.2	99.8	2.88
10	90	1.28	0.1	99.9	3.09
5	95	1.64	0.05	99.95	3.29
2.5	97.5	1.96	0.02	99.98	3.54
2.3	97.7	2.00	0.01	99.99	3.72
2	98	2.05	0.005	99.995	3.89
1	99	2.33	0.002	99.998	4.11
0.5	99.5	2.58	0.001	99.999	4.27

For simple cases one can often write

$$S_d = \gamma_{Sd} \gamma_f S_k \qquad [15.7]$$

and

$$R_d = \frac{R_k}{\gamma_m \gamma_{Rd}} \qquad [15.8]$$

so the acceptance criterion becomes

$$S_k \leq \alpha R_k \qquad [15.9]$$

where

$$\alpha = \frac{1}{\gamma_m \gamma_f \gamma_{Sd} \gamma_{Rd} \gamma_n} \qquad [15.10]$$

If S is a stress, this corresponds to the familiar stress factor (or allowable stress) format. A load factor format is similarly obtained if the γ factors are instead combined into one or more factors applied to the loads F_k.

In the level I approach the partial factors may be determined on the basis of experience and judgement, or by means of a higher level probabilistic evaluation that aims to ensure a specified target reliability. The γ factors will then take on typical values that correspond to a particular target reliability level for a family of generic structural types, rather than saying anything about the reliability (i.e. probability of survival) of the individual structure.

Even when based on limit state principles with partial safety factors determined by higher level probabilistic methods, the level I approach concentrates on individual limit states associated with individual failure modes, often in individual elements or connections in the structure. It does not provide a way of assessing quantitatively the total safety taking account of all possible loading conditions and failure modes.

Level II – Semi-probabilistic methods

The level II approach is semi-probabilistic in that it takes account of more statistical information than a single, characteristic value for each variable. More specifically, each stochastic variable is represented by two parameters, usually the expected value (i.e. the mean) and the variance. In addition the correlations between variables are taken into account, usually through the covariances. These correlations are mainly relevant to the applied loads. Level II methods, when the expected values, variances and covariances are used, are also known as **second moment** methods.

In a level II approach one aims to characterise the reliability of a given structure or component, subjected to a given loading, by a single numerical parameter, referred to as the **safety index**, or **reliability index**.

We represent the basic variables by lower case z_i when treated as deterministic variables and by upper case Z_i when treated as stochastic variables. The expected (mean) values are written $E[Z_i]$, the covariances $Cov[Z_i,Z_j]$ and the variances $Var[Z_i]$. Note that the variances are included in the covariances.

Suppose that analysis of the structure indicates that, for the limit state under consideration, failure occurs when a function $g(z_i)$ (referred to as the **limit state function**) is zero, in such a way that

$$g(z_i) > 0 \qquad [15.11]$$

represents safe states of the structure and

$$g(z_i) < 0 \qquad [15.12]$$

represents unsafe (failed) states. Then

$$g(z_i) = 0 \qquad [15.13]$$

defines a surface which can be referred to as the **failure surface**.

In many cases the function g can be written in the form

$$g(r,s) = r - s \quad [15.14]$$

where r and s are deterministic variables representing resistance and load effect (symbols R and S now being reserved for the corresponding stochastic variables).

We define a **safety margin** for the case with random variables:

$$M = g(Z_i) \quad [15.15]$$

Cornell [8] defined a reliability index (or safety index)

$$\beta_C = \frac{E[M]}{D[M]} \quad [15.16]$$

E and D are the mean and standard deviation of M.

This is illustrated schematically in Figure 15.2. If M is normally distributed, Table 15.1 (in which the β values now correspond to β_C) shows that $\beta_C = 1$ corresponds to a 16% failure probability, while $\beta_C = 2$ gives 2.3%, $\beta_C = 3$ gives about 0.15% and $\beta_C = 4$ gives about 0.003% i.e. 3×10^{-5}, and so on.

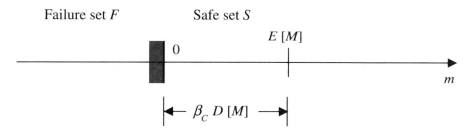

Figure 15.2 *Illustration of the Cornell reliability index* β_C.

If g is a linear function of the basic variables such that

$$g(z_i) = a_0 + \sum_{i=1}^{n} a_i z_i \quad [15.17]$$

and C_{ij} are the covariances, it is possible to show that

$$\beta_C = \frac{a_0 + \sum_{i=1}^{n} a_i E[Z_i]}{\sqrt{\sum_{i=1}^{n} a_i a_j C_{ij}}} \quad [15.18]$$

For cases where equation [15.14] applies,

$$M = R - S \quad [15.19]$$

Then, if R and S are uncorrelated, we obtain the special form

$$\beta_C = \frac{E[R] - E[SS]}{\sqrt{Var[R] + Var[S]}} = \frac{\mu_R - \mu_S}{\sqrt{\sigma_R^2 + \sigma_S^2}} \qquad [15.20]$$

μ and σ represent mean and standard deviation values.

From these equations it is possible to derive analytical formulae for β_C for specific structural cases. This and similar approaches work satisfactorily when the failure surface is a hyperplane, i.e. when the limit state function g is a linear function of the basic variables z_i (so M is a linear function of the stochastic variables Z_i), but it cannot be used directly when a nonlinear function is involved. In such cases it is necessary to approximate the limit state function using a linear expansion about a selected point. One possibility is to use the linear terms of a Taylor series expansion about the point representing the expected values of the stochastic variables. When the limit state function is linearised the resulting index is referred to as a **first order, second moment reliability index**.

Note that, for a given problem, the function $g(z_i)$ is not unique although the failure surface itself may be unique. There are some problems in defining a reliability index such that it is not dependent on the choice of $g(z_i)$ for a particular failure surface. The first order index is also dependent on the point selected for the linear expansion.

Hasofer and Lind[9] proposed the following procedure to overcome these problems of uniqueness. The reliability index defined earlier can be interpreted as a measure of the distance to the failure surface from the point representing the expected values. In the one-dimensional case the standard deviation of the safety margin was used as a scale (see Figure 15.2). When there are two or more basic variables it is possible to find a nonhomogeneous, linear transformation of the basic stochastic variables Z_i into a set of normalised and uncorrelated variables X_i, i.e. such that

$$E[X_i] = 0 \qquad [15.21]$$

and with unit covariance matrix

$$C_X = Cov[X, X^T] = I \qquad [15.22]$$

This transformation requires standard linear algebra techniques using eigenvalues/eigenvectors. Thus the mean-value point in Z–space is mapped into the origin in the X–space, and the failure surface L_Z mapped into a new surface L_X (see Figure 15.3). The geometrical distance from the mean point in X–space to a given point on L_X is the number of standard deviations from the mean point in Z–space to the corresponding point on L_Z. The reliability index β_{HL} is then taken as the *smallest* such value, which in X–space is the *smallest distance* from the origin O to the failure surface. The problem then becomes one of finding the location of the **design point**, which is defined as the point on the failure surface L_X that lies nearest to O. This can be solved using standard optimisation techniques. Further generalisations of this idea are possible and may be needed in cases where the failure surface has particular geometric features such as high curvature.

In principle the level II reliability approach can be used to calculate reliability indices directly in a design situation. However, it has mainly been used to establish numerical values of partial factors for use in a level I code formulation. Nowadays automated computer methods are available that enable one to take account of more information about the stochastic variables than just the means and the covariances. These are described in the next section.

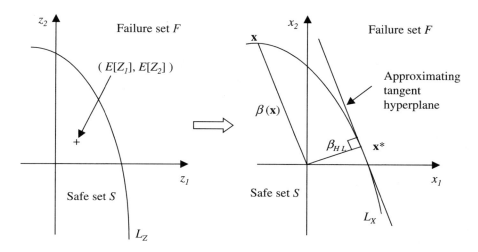

Figure 15.3 *Illustration of the Hasofer and Lind reliability index* β_{HL}.

Level III – Probabilistic methods

In a level III method a fully probabilistic analysis is performed in which all possible combinations of loads and strengths are explored to establish the probability that a given limit state will be exceeded. For this approach one needs relatively detailed information about the statistical variation of each stochastic variable and also the correlations between them. These methods apply the probability of failure P_F as a basic measure:

$$P_F = \int_F f_Z(z)dz = P(M < 0) \qquad [15.23]$$

$f_Z(z)$ = the joint probability density function of the basic variables
F = the failure set

The probability of failure may be transformed into a generalised reliability index β_R, which is defined as

$$\beta_R = \Phi^{-1}(P_R) = -\Phi^{-1}(P_F) \qquad [15.24]$$

in which Φ is the cumulative distribution function for a normally distributed variable with zero mean and unit variance. Thus β_R is the argument of the normal distribution function at the point corresponding to the survival probability P_R, or minus that at the failure probability P_F.

There are various ways of calculating, or estimating, β_R. If the stochastic variables are jointly normally distributed, and the failure surface is a hyperplane, the second moment index approach of level II semi-probalistic methods gives the result directly. If the failure surface is not a hyperplane, the Hasofer and Lind first order, second moment approach often gives a good approximation. This applies provided the design point represents the only local minimum of the distance from the origin to the failure surface in X–space. The reason is that the probability density function of the basic variables decreases very rapidly with the distance from the origin. The area of integration contributing most to the failure probability is thus close to the design point, and in this region the tangent hyperplane around the design point is a good approximation to the failure surface.

In general the basic variables are not normally distributed, but it is possible to perform a further transformation from the uncorrelated variables X_i to a new set of variables U_i that are normally distributed. The limit state surface in x–space is then mapped into a corresponding limit state surface in u–space. The design point is now determined in u–space by finding the point on the failure surface that is nearest to the origin. The failure probability is then approximated to a first order by finding $\Phi(-\beta)$, where β is the distance from the origin to the design point. The method is then referred to as a **first order reliability method (FORM)** and β is the **first order reliability index**.

Better approximations can be obtained by using other approximate representations of the failure surface such as quadratic or multi-plane approximations. Use of a quadratic approximation at the design point leads to a **second order reliability method (SORM)**. An exact integral for the reliability index has been established for this case by Tvedt.[10]

Computer programs are available to calculate structural reliability using level III methods. An example is PROBAN.[11,12] This gives several options as regards the actual method of calculation used:

- First order approximation (FORM)
- Second order approximation (SORM)
- Monte Carlo and other simulation methods

Level III methods are often used to assess individual offshore structures, but otherwise it is rarely feasible for individual structures because of the effort involved and the need for so much statistical data. However, the level III approach is commonly used for calibrating level I methods, i.e. for determining the partial safety factors to be used in these approaches for a given class of problems.

An advantage of level III methods is that they open for the calculation of **system reliability**. Instead of only considering a single failure mode under a given condition of use of the structure, it is possible to combine the failure probabilities of many modes of failure and conditions of use and loading without introducing

Level IV – reliability methods

Level IV methods are basically similar to level III, but the analysis goes further in that it considers not only the probability aspect but also the *consequence of failure*, in terms of, for example, cost to the owner of the structure or to society in general. This allows, for example, the determination of **target reliability levels** that should be adopted for lower level methods. For example, in a level I code it will help to decide how the factor γ_n should vary according to type of structure and type of limit state.

15.4 TARGET RELIABILITY LEVELS

Some typical target annual failure probabilities and reliability levels for structural design, taken from a Nordic building standard, are shown in Table 15.2.

Table 15.2 Target annual failure probabilities and corresponding reliability indices, from NKB[13]

Failure development	Failure consequences		
	Not serious	Serious	Very serious
Ductile failure with reserve strength capacity	$P = 10^{-3}$ $\beta = 3.09$	$P = 10^{-4}$ $\beta = 3.71$	$P = 10^{-5}$ $\beta = 4.26$
Ductile failure with no reserve capacity	$P = 10^{-4}$ $\beta = 3.71$	$P = 10^{-5}$ $\beta = 4.26$	$P = 10^{-6}$ $\beta = 4.75$
Brittle failure in terms of fracture or instability	$P = 10^{-5}$ $\beta = 4.26$	$P = 10^{-6}$ $\beta = 4.75$	$P = 10^{-7}$ $\beta = 5.20$

15.5 SOME TYPICAL DISTRIBUTIONS

Table 15.3 shows which distributions are found to apply for a number of stochastic variables that are relevant to marine structures. Some of these are also relevant to sailing yachts.

Table 15.3 Distribution types for some standard variables, from DNV[3]

Variable name		Distribution type
Wind	Short-term instantaneous gust speed	Normal

Waves	Long-term n-minute average speed	Weibull
	Extreme speed, yearly	Gumbel
	Short-term instantaneous surface elevation (deep water)	Normal
	Short-term heights	Rayleigh
	Wave period	Longuet-Higgens
	Long-term significant wave height	Weibull
	Long-term mean zero upcrossing or peak period	Log-normal
	Joint significant height / mean zero upcrossing or peak period	3-parameter Wiebull (height) / Log-normal period conditioned on ht.
	Extreme height, yearly	Gumbel
Forces	Hydrodynamic coefficients	Log-normal
Fatigue	Scale parameter of S–N curve	Log-normal
	Fatigue threshold	Log-normal
Fracture mechanics	Scale parameter of da/dN curve	Log-normal
	Initial crack size	Exponential
	P.O.D. curve	Log-normal or Weibull
Properties	Yield strength (steel)	Log-normal
	Young's modulus	Normal
	Initial deformation of panel	Normal
Ship data	Still water bending moment	Normal
	Joint still water moment / draught	Joint normal
	Ship speed	Log-normal
	Model uncertainty of linear calculations	Normal

Material strength properties can usually be represented by either a normal or a log-normal distribution. The advantage with the log-normal distribution is that it does not include negative values. For **FRP composites** the fibre strengths are generally found to follow a Weibull distribution. However, when the fibres are combined with a matrix to form a laminate, the resulting strength is close to normally distributed.

15.6 MATERIAL FACTORS FOR LEVEL I DESIGN: FIBRE REINFORCED COMPOSITES

Individual partial safety factors for level I design of structures are sometimes expressed as the product of several component factors that are chosen to cover different uncertainty aspects. For ultimate limit states of structures built in fibre reinforced plastics the EUROCOMP Design Code[14] expresses the partial safety factor for material properties in the form

$$\gamma_m = \gamma_{m,1} \gamma_{m,2} \gamma_{m,3} \qquad [15.25]$$

$\gamma_{m,1}$ account for the level of uncertainty relating to derivation of material properties from test values
$\gamma_{m,2}$ account for uncertainties of the material and production process
$\gamma_{m,3}$ account for uncertainties regarding environmental effects and duration of loading

The suggested values of these factors are presented in Tables 15.4–15.6.

Table 15.4 Factor $\gamma_{m,1}$ to cover uncertainties regarding derivation of material properties

Derivation of properties	Value of $\gamma_{m,1}$
Properties of constituents are derived from test specimen data, properties of individual laminae are derived from theory and properties of the laminate or panel are derived from theory	2.25
Properties of individual plies are derived from test specimen data and properties of the laminate or panel are derived from theory	1.5
Properties of the laminate or panel are derived from test specimen data	1.15

Table 15.5 Factor $\gamma_{m,2}$ to cover uncertainties regarding material and production process

Production method	Value of $\gamma_{m,2}$	
	Curing: Fully postcured	Not fully postcured
Hand-held spray	2.2	3.2
Machine-controlled spray	1.4	2.0
Hand lay-up	1.4	2.0
Resin transfer moulding	1.2	1.7
Pre-preg lay-up	1.1	1.7
Machine controlled filament winding	1.1	1.7
Pultrusion	1.1	1.7

Table 15.6 Factor $\gamma_{m,3}$ to cover uncertainties regarding environmental effects and duration of loading

Operating design temperature (°C)	Heat distortion temperature (HDT) (°C)	Value of $\gamma_{m,3}$	
		Short-term loading	Long-term loading
25–50	55–80	1.2	3.0
	80–90	1.1	2.8
	>90	1.0	2.5
0–25	55–70	1.1	2.7
	7–80	1.0	2.6
	>80	1.0	2.5

Tables like these are helpful in that they differentiate between the sources of uncertainty in strength values, and give credit both for more reliable estimates of strength and for the lower variability associated with better controlled production processes. However, if they are to be of real use it is important

1. To define explicitly which fractile the characteristic values of the material properties are assumed to be based on (whether these are measured at the level of constituent materials, plies or laminates), and
2. To provide some means of accounting for the number of measurements from which the properties have been determined.

These are both crucial points because the optimal value of the partial factor γ_m depends on how much uncertainty has already been covered in the selection of the characteristic value. The second aspect is discussed more fully below.

For marine composites, a particular material strength property is often found from tests on a rather limited number of specimens (typically about 5). This leads to an appreciable uncertainty in the estimates of the mean and standard deviation, and hence of the given fractile used for the characteristic value. If we denote the estimated value of the given fractile, based on N measurements, by r_N, and the 'true' value (that would come from an infinitely large number of measurements) by r_c, there is a roughly equal probability that r_c will be greater or smaller than r_N. However, it is possible to adjust the estimate so that there is a higher probability (called the confidence level) that $r_N < r_c$, i.e. that the estimate is on the safe side. A mathematical description of this is rather complicated, but for a normally distributed variable it is possible to tabulate the number of estimated standard deviations that must be subtracted from the estimated mean to obtain a given fractile with a given confidence level for a given sample size. As an example, to obtain the true 2.5% fractile on the basis of an infinite sample size requires subtraction of 1.96 standard deviations, as shown in Table 15.1. However, with estimates based on 10 test specimens, to obtain the same fractile with 90% confidence requires subtraction of 3.01 estimated standard deviations, and for only

5 specimens it requires subtraction of 3.98 standard deviations. Thus the characteristic value should be specified not only in terms of the fractile of the distribution, but also in terms of either the number of measurements or the confidence level.

15.7 METHODS FOR ASSESSING OVERALL SAFETY

Structural reliability analysis is normally applied separately to individual limit states corresponding to individual failure modes. However, using a level III approach it is possible to combine the probabilities of the separate failure modes into a single overall probability measure for failure of the structure as a whole. This is referred to as a **system reliability** approach. At present such methods are in their infancy, but advanced computer tools such as PROBAN[11,12] are able to deal with such problems.

Note that the calculation of system probability is not in general a simple matter of summing the probabilities for the separate failure modes, because the stochastic variables occurring in the separate modes are not in general statistically independent.

15.8 QUANTITATIVE RISK ANALYSIS AND FORMAL SAFETY ASSESSMENT

Many loads are relatively well known and can be described quite well in statistical terms. However, some loads occur only under very exceptional conditions, often as a consequence of human error or an unforeseen situation. Examples are accidental loads due to collisions between ships or between a ship and a fixed object such as a quay or a rock. Other examples are the loadings due to fires and explosions. These loadings require a different type of description from loads such as hydrodynamic loads and motion-related inertia loads.

Two alternative approaches might be envisaged for coping with loads of this type:

- Take measures to ensure that the probability of occurrence of the event in question is low enough to be considered acceptable
- Take measures to ensure that the structure can withstand such an event without damage, or with an acceptable, limited extent of damage

In deciding on the appropriate measures it is necessary to consider not only the probability of occurrence but also the severity of the consequences if such an event does occur. This leads to the concept of **risk**:

Risk may be defined as the combination of the probability of an unfavourable event and the severity of its consequences.

In risk analysis the probability aspect is most often represented by the **frequency** with which such events occur, or are expected to occur. Figure 15.4 illustrates the trade-off between frequency (or probability) and consequence. It may be necessary to consider risks to people, property or the environment.

Risk analysis consists in general of the following steps:

- Identification of hazards
- Assessment of probabilities/frequencies of initiating events
- Assessment of how an initiating event can develop into different accidental events
- Assessment of consequences of different accidents
- Calculation of risks

The term **risk assessment** is often used to include both a risk analysis and an evaluation of the results to make decisions.

There are established techniques for identifying hazards and for quantifying the risks associated with them. In particular should be mentioned

- fault trees, which show the causal relationship between events which singly or in combination result in the occurrence of a higher level event, where the highest level event may be a type of accident or unintended hazardous outcome, and
- event trees, which show the sequence of events arising from an accident, a failure or an unintended event

Fault trees help in identifying causes of accidents and ways of reducing their probability of occurrence. Event trees help in identifying the consequences and ways of mitigating them.

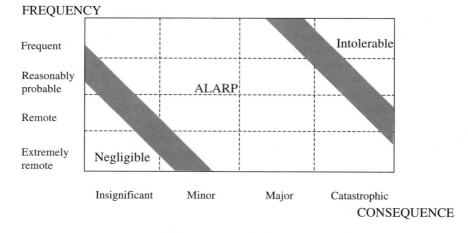

Figure 15.4 *Graphical illustration of the risk matrix. ALARP = as low as reasonably practicable. The risk level boundaries shown are purely illustrative.*

One way of approaching accidental loads is to require that a structure shall be capable of surviving a particular accident scenario. Examples applied to building structures are gas explosions of specified intensity and extent, and collisions of specified vehicles with the structure at specified speed. Examples for ships include the requirement in the current IMO Code of Safety for High Speed Craft[15] for the hull structure to withstand a specified collision scenario (collision at service speed with a rock or breakwater protruding 2 m above the water surface) without exceeding a specified deceleration level or suffering damage to the passenger compartment. However, unless the scenarios are selected in a particularly skillful way, such approaches to accidental loadings can result in very inconsistent levels of safety for different types or dimensions of structure or ship.

The use of **formal safety assessment (FSA)**[16] helps to remove the arbitrariness from scenario selection and to achieve a level of safety that is both consistent and cost effective. FSA is applied to a class of structures or to a generic type of ship, rather than to a single, specific structure or ship.

An FSA consists of the following steps:

1. Identification of hazards.
2. Assessment of risk associated with these hazards.
3. Consideration of options for controlling the identified risks.
4. Cost/benefit assessment of these options.
5. Recommendations for decision making.

These are not necessarily to be taken in the above sequence; the process is shown schematically in Figure 15.5.

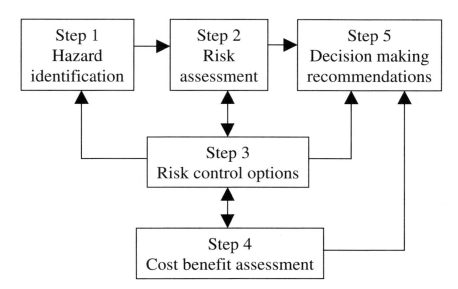

Figure 15.5 *Flow chart illustrating Formal Safety Assessment (FSA) methodology.*

FSA has so far been applied in very few cases related to ships. An example in which FSA techniques (though not all aspects of an FSA) have been applied is that of bulk carriers, which are a ship type that has involved the loss of more than 600 lives in a period of 17 years. An analysis performed by DNV[17] showed that risk associated with vessels of this type has been partly in the intolerable region. Options for reducing the risks for *existing ships* that were either already implemented or proposed by the International Association of Classification Societies included an enhanced survey to increase the probability of detection of corrosion and fatigue damage, and a requirement for strengthening of the bulkhead between the first and second cargo holds in such a way that it would tolerate flooding of the first hold without collapsing. The first of these measures was found to reduce the risk appreciably, and when combined with the second measure the risk was reduced well into the ALARP region. Furthermore, the cost/benefit analysis of the bulkhead strengthening requirement showed that, when compared with requirements in other sectors, this measure is cost-effective.

FSA normally assumes the availability of a substantial amount of data regarding the family of problems being considered, whether it is a family of physical applications (e.g. a generic type of ship) or a family of event types (e.g. a generic type of accident such as collision or fire). The technique is still in its infancy but the IMO has recently issued draft guidelines for its use in the rule-making process.[16]

15.9 APPLICATION OF SAFETY ENGINEERING PRINCIPLES TO SAILING YACHTS

At present the structural design of smaller, conventional sailing yachts used for pleasure purposes is largely based on relatively simple standards such as the Nordic Boat Standard (which applies to pleasure boats under 15 m in length). Otherwise, and particularly for offshore racing yachts, the rules of classification societies are used, such as the DNV Rules for Classification of High Speed and Light Craft,[18] the Lloyd's Register Rules and Regulations for the Classification of Special Service Craft,[19] or the ABS Guide for Building and Classing Offshore Racing Yachts.[20] Such rules and guidelines make little, if any, use of risk- and reliability-based approaches at present. Simple, allowable stress formats, combined with prescriptive requirements based mainly on experience, are most common. Risk- and reliability-based approaches are, however, finding their way into the design of larger ships and some aspects such as formal safety assessment are beginning to find application in the context of high-speed ferry safety. Such approaches provide considerable scope for application to smaller craft in the future, both in terms of development of improved rules, guidelines and international regulations and in relation to individual projects such as high-performance craft for which investment in advanced design methods is not just possible but also cost effective.

REFERENCES

1. ISO 2394: General principles on reliability for structures, 2nd edition, 1986. International Organization for Standardization. (Under revision).
2. ENV 1991-1 Eurocode 1 – Basis of design and actions on structures – Part 1: Basis of design. European prestandard, Sept. 1994, European Committee for Standardization.
3. Classification Note CN 30.6 'Structural reliability analysis of marine structures'. Det Norske Veritas, Norway, 1992.
4. Madsen, H.O., Krenk, S. and Lind, N.C. *'Methods of Structural Safety'*. Prentice-Hall, Inc., Englewood Cliffs, NJ 07632, USA, 1986.
5. Ditlevsen, O. and Madsen, H.O., *'Structural Reliability Methods'*. John Wiley & Sons Ltd., Chichester, England, 1996.
6. Thoft–Christensen, P. and Baker, M.J., *'Structural Reliability and its Applications'*. Springer–Verlag, Germany, 1982.
7. Skjong, R. *et al.* *'Guideline for offshore structural reliability analysis – General'*. DNV Report No. 95–2018, Det Norske Veritas, Norway, 1996.
8. Cornell, C.A. *'A Probability-Based Structural Code'* Journal of the American Concrete Institute, Vol. 66, No. 12, 1969, pp. 974–985.
9. Hasofer, A.M. and Lind, N.C., *'Exact and Invariant Second-Moment Code Format'*. Journal of Engineering Mechanics Division, ASCE, Vol. 100, 1974, pp. 111–121.
10. Tvedt, L. *'Second order reliability by an exact integral'*. Proceedings, 2nd IFIP Working Conference on Reliability and Optimisation of Structural Systems, ed. P. Thoft–Christensen, Springer Verlag, Germany, 1993.
11. Tvedt, L. *'PROBAN Version 4 Theory Manual'*, DNV Research Report No. 93–2056, Det Norske Veritas, Norway, 1993.
12. Det Norske Veritas. *'PROBAN User's Manual'*. DNV Report No. 92–7049, Rev. 1, Det Norske Veritas, Norway, 1996.
13. NKB. *'Regulations for Loading and Safety Regulations for Structural Design'*. NKB Report No. 36, Nordic Committee on Building Regulations, Copenhagen, Denmark, Nov. 1978.
14. Clarke, J.L. (ed.) *'Structural Design of Polymer Composites'* (also known as EUROCOMP Design Code and Handbook). E & F.N. Spon, London, U.K., 1996.
15. *'Code of Safety for High Speed Craft'*. International Maritime Organization, 1995.
16. *'Interim Guideline for the Application of Formal Safety Assessment (FSA) to the IMO Rule Making Process'*. International Maritime Organization, 1997.
17. Det Norske Veritas. *'Cost Benefit Analysis of Existing Bulk Carriers – A Case Study on Application of Formal Safety Assessment Techniques'*. DNV Paper Series No. 97–P008, Det Norske Veritas, Norway, 1997.
18. Det Norske Veritas. *'Rules for Classification of High Speed and Light Craft'*. Det Norske Veritas, Norway, 1991–96.
19. Lloyd's Register. *'Rules and Regulations for the Classification of Special Service Craft'* Lloyd's Register of Shipping, U.K., 1996.
20. American Bureau of Shipping. *'Guide for Building and Classing Offshore Racing Yachts'*. American Bureau of Shipping, USA, 1994.

INDEX

accommodation 203
American Bureau of Shipping (ABS) 168, 277, 331
apparent wind 3, 36, 109, 301
aspect ratio 33, 59, 107, 161, 172, 197

backstay 193
balance
  force 7, 113
  hull–rig 73, 104
  laminate 160, 272
ballast 10, 17, 111, 149, 283
beam 16, 47, 91, 112, 169, 194, 218, 258, 283
beam to draft ratio 69, 91, 118
beam to length ratio 25
beams 166, 194, 275
bending moment 172, 194, 271, 315
biplane theory 115
BM 16
BOC 58
body plan 216
boom 41, 193
boundary layer 5, 28, 46, 127, 236, 281
break angle 138
breaking waves 101
brittle fracture 148
buckling 145, 186, 194, 259, 312
bulb 65, 117, 177, 254, 282
bulkhead 23, 169, 201, 330

buoyancy 8, 14, 57
Bureau Veritas 171, 206

camber 30, 119, 229, 283
carbon fibre 181, 197
catamaran 17, 137
cavitation 235
centre of
  buoyancy 14, 57
  effort (CE) 6, 43, 104, 130, 285
  flotation 57
  gravity 10, 14, 65, 88, 112, 191
  lateral resistance (CLR) 6, 123, 285
  pressure 124
chainplates 192, 258
classification societies 2, 171, 213, 259, 331
climate 124
cockpit 20
collision 148, 310
composite 26, 147, 190, 197, 274
computational fluid dynamics (CFD) 2, 52, 111, 232, 235, 278
computer-aided design (CAD) 216, 260
computers 191, 235, 260, 278
control surfaces 241
core 138, 149, 179, 259
corrosion 151, 165, 196, 330
cracking 176, 259, 314
critical load 201
curvature 218

333

damage stability 26
damping 88, 134
Delft Systematic Yacht Hull
 Series 50
Delft University of Technology 46, 78, 190
density
 fluid 147, 179, 237
 probability 311
depth 61, 118, 160, 175, 258
design spiral 228
diagonal 292
discontinuities 176, 216
displacement 11, 14, 57, 86, 117, 169, 194, 243, 261, 289, 315
downflooding 25
draft 8, 47, 91, 117, 205, 282, 324, 330
 effective 69, 121, 296
drag 2, 4, 27, 46, 113, 237, 293
 bucket 33
dynamic stability 104

equilibrium conditions 10, 124
EU Recreational Craft Directive 171
Euler 254

failure 146, 176, 275, 312
fairing 115, 216
fatigue 88, 145, 165, 203, 259, 314
fetch 80, 124
fibre content 179
fibre reinforced plastics (FRP) 149, 165, 324
finite element analysis (FEA) 187, 196, 258
flare 22
flooding 25, 330
flotation 18, 57, 111
flutter 39
forestay 193
fracture 146, 323
frameworks 312
free surface 22, 47, 138, 233, 242
freeboard 19, 293

frequency of encounter 100
frictional resistance 46, 117
Froude number 54, 119, 242

girder 2, 154, 176, 258
grain 155
grounding 177, 259
GZ curve 15, 123

heave 78, 286
heel 15, 41, 109, 204, 232, 285
 equilibrium 8
 resistance 61, 97
hogging 173
hydraulic 201
hydrostatic pressure 175
hydrostatics 135, 215

impact 147, 178, 259
induced resistance 46, 122
ingress of water 26
initial stability 15
instrumentation 312
integration 119, 232, 240, 269, 322
International America's Cup Class (IACC) 59, 143, 252, 282
International Measurement System (IMS) 56, 96, 10, 232, 285
International Offshore Rule (IOR) 41, 110
International Rule
 6 m 194, 303
 8 m 151, 279
 12 m 96, 194, 279
International Standards Organisation (ISO) 153, 171, 314
International Towing Tank Conference (ITTC) 49, 117, 308
inverse taper 67

keel 7, 17, 27, 47, 105, 115, 146, 173, 199, 216, 238, 259, 290
 bolts 259

Index 335

laminar flow 29, 49, 243, 290
laminate 157, 166, 325
large angle stability 24
lead 101
leeway 7, 123, 295
length
  effective 50, 119
  IMS 118
  overall 12, 91
  panel 195
  scantling 171
  waterline 56, 205
  wave 79
length to displacement ratio 86
lift
  aerodynamic 5, 28, 127
  hydrodynamic 47, 78, 111
lifting line 237
lines plan 60, 215
load factor 312
local loads 186
longitudinal centre of buoyancy (LCB) 58, 118, 231
longitudinal centre of gravity (LCG) 24

mainsheet 203
manoeuvring 101
Massachusetts Institute of Technology (MIT) 76, 109, 257
mast 3, 30, 112, 174, 197, 243, 303
  step 258
mean 5, 34, 89, 118, 146, 174, 229, 245, 259, 281, 316
midship section 216
model tests 55, 83, 213, 235, 273, 284
moment of inertia 90, 173, 195
mylar 304

NACA 31, 52, 229, 282
neutral axis 154, 176
Nordic Boat Standard 331
notch toughness 151
NURBS 224

offsets 229

performance analysis 110
pitch 4, 25, 78, 191, 286
planform 34, 63, 227, 282
plasticity 271
potential flow 46, 236
prepreg 149
pressure coefficient 251
prismatic coefficient (CP) 58, 90
profile 34, 52, 131, 154, 176, 199, 216, 243, 306
  diagonals 230
Prohaska 51, 292
propulsion 192

range of stability 15
Registro Italiano 190
reliability method 322
residuary resistance 56, 117, 285
resistance 8, 18, 55, 116, 124, 242, 289
  added in waves 78, 94, 115, 298
  components 47, 101, 116, 286
  impact 145
  upright 55, 109, 283
response amplitude operator (RAO) 82
Reynolds number 49, 144, 236
rig 10, 18, 36, 109, 172, 193, 252, 259, 296
  types 143, 308
rigging 35, 88, 127, 147, 173, 192, 259, 303
righting moment 8, 14, 112, 205
roll 4, 78, 135, 191, 238, 289
Royal Institution of Naval Architects (RINA) 45, 77, 143, 162, 171, 308
rudder 3, 47, 101, 115, 238
  area 284
  stock 173

safety engineering 310
safety factors 146, 170, 313
sagging 147, 173

sail forces 27
sand strip 290
sandwich 26, 155, 166, 259
scale factor 231
scaling 290
sea surface 38, 243, 305
seakeeping 90, 284
second moment of area 16, 117
section modulus 173, 207
sectional area 59, 118, 261
separation 8, 14, 29, 71, 127, 243, 298
shear 153, 179, 216, 243, 259
  force 315
sheer 293
ship motion 72, 82
shroud 173, 193
  angle 193
  tang 202
sideforce 120
  sailing 9, 116
significant wave height 94, 324
simulation 214, 301, 322
skin friction 52, 243
slamming 169, 260, 313
  loads 169, 313
Society of Naval Architects and Marine Engineers (SNAME) 45, 76, 108, 143, 166, 214, 233, 308
specific gravity 151
spectra 84, 298
spinnaker 42, 115, 178
spline 218, 295
spreader 193, 304
stability 8, 14, 65, 101, 173, 198, 232, 296
  assesment 107
  dynamic 104
stall 30, 71, 107, 130, 295
stiffness 15, 145, 164, 264, 312
strain 146, 185, 218, 246, 261, 286, 315
  energy 148, 218, 266
strength 156, 259, 315
stress 146, 170, 186, 206, 244, 251, 261, 275

strip theory 89
structural response 183
struts 298
superstructures 176, 232
surface 222, 243
surge 82, 287
sway 82, 287
sweep angle 64, 229

tandem keel 284
taper ratio 52
temperature 148, 289, 326
thrust 99, 113, 295
torque 135, 178, 315
towing tank tests 285
transition 29, 49, 243, 290
transom 230, 281
trim 26, 62
  surfaces 227
  tab 299
true wind 3, 78, 110, 302
turbulent flow 28
twist 37, 138, 222, 301

ultimate strength 146, 185, 314
University of Southampton 282

variance 318
velocity
  prediction program (VPP) 2, 92, 109, 283
  triangle 7, 305
ventilation 25, 74
vertex 219
vertical centre of
  buoyancy (VCB) 26
  gravity (VCG) 10, 23, 115
viscosity 28, 46, 236
viscous resistance 48, 117, 281
volume 21, 47, 117, 148, 242, 258, 293
vortex 33, 64, 122, 197, 237

waterline length 47, 88, 118, 205, 281

waterlines   90, 118, 216
waterplane   3, 16, 68, 121
wave   3, 25, 46, 78, 115, 148, 173, 280, 312
  frequency   82
  pattern   53, 123, 296
  resistance   46, 91, 119, 242
  sinusoidal   89
  spectrum   79, 91
weight   8, 145
  distribution   16
wetted surface area   117

wind   3, 18, 27, 48, 78, 109, 192, 279, 312
  speed   5, 80, 110, 279
  strength   38, 80, 127, 307
wind tunnel tests
  appendages   122, 298
  sails   42, 132, 301
windage   35, 115, 191, 304
wing section   52
winglet   122, 280

yaw   4, 82, 113, 254, 285